T0402275

Laser Lithotripsy

R. Steiner (Ed.)

Laser Lithotripsy

Clinical Use and Technical Aspects

With 129 Figures

Springer-Verlag Berlin Heidelberg NewYork
London Paris Tokyo

Professor Dr. Rudolf W. Steiner
Institut für Lasertechnologien in der Medizin, Universität Ulm,
Postfach 40 66, D-7900 Ulm, Fed. Rep. of Germany

ISBN 978-3-642-73866-1 ISBN 978-3-642-73864-7 (eBook)
DOI 10.1007/978-3-642-73864-7

Softcover reprint of the hardcover 1st edition 1988

Printing: Druckhaus Beltz, 6944 Hemsbach/Bergstr.
Binding: J. Schäffer GmbH & Co. KG., D-6718 Grünstadt
2156/3150-543210 – Printed on acid-free paper

Preface

Laser lithotripsy is a new field of laser applications in medicine, of growing importance for the fragmentation of stones in the human body. This new technique does not necessarily compete with the established extracorporal lithotripsy but rather extends the applications to include the disintegration of stones (kidney, ureter and gall stones) which cannot be fragmented by other means. In combination with endoscopic techniques, the advantages of laser lithotripsy are the fragmentation of stones into extremely small particles, or even dust, and the possibility of rinsing out the debris. This is certainly an advantage in stone disease therapy. The low costs of laser systems – in contrast to extracorporal systems – will promote the spread of this promising techniques.

There are, however, still some questions requiring discussion: Which laser should be used? Which wavelength is the most efficient? And what pulse duration for high energy transmission through thin quartz fibres can be tolerated? Laser and fibre-optic technologies are essential in the development of an effective medical laser system. Their specific features must be based on fundamental research of the laser-stone interaction and also practicability in clinical use.

This book consists of papers presented at the 1st International Symposium on Laser Lithotripsy and provides a summary of the current state of the art in basic research into laser reactions with stones, instrumental development and clinical applications. It also does not neglect possible tissue damage by this method, which in any case is negligible compared to extracorporal shockwaves. Shock waves penetrating the body may be painful, the amount of pain depending on the form of the pressure amplitudes, especially the amount of the negative component. Shock waves produced by lasers directly at the stone surface are free from this negative influence.

This volume gives a complete overview of the laser technology relevant to laser lithotripsy, laser transmission systems, basic reactions, pressure measurements and clinical applications. These results will certainly stimulate further developments towards the "ideal" laser system for stone fragmentation.

Ulm, February 1988 *R. Steiner*

Contents

History of the Management of Stone Disease

R. Hautmann

University of Ulm, Department of Urology, Prittwitzstraße 43,
D-7900 Ulm, Fed. Rep. of Germany

The history of urolithiasis apparently is as old as mankind. It is not only a long story but it is also a fascinating document which records on the one hand the stoicism of the patients who over many hundreds of years submitted themselves to the agony of surgery in order to escape the tortures of the stone and on the other hand, the quite amazing ingenuity and courage of the surgeons who developed and perfected the techniques of cutting for the stone or removing its crushed fragments.

The oldest stone so far discovered was obtained from the grave of a mummy of a boy aged about 16 years in the pre-historic cemetery at El Amrah in Upper-Egypt and was dated at about 4800 B. C. (Fig. 1).

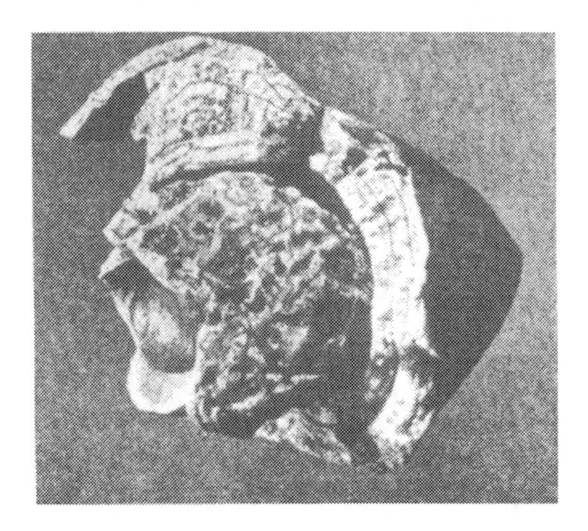

Fig. 1

Of the triad of elective operations first performed by man - circumcision, trephination of the skull and cutting for bladder stone - the last was the only one free from religious or ritual conventions and may therefore safely be pronounced the most ancient operation undertaken for the relief of a specific surgical condition.

Descriptions of means to relief the patient of his stone has come down to us by ancient writings of church and medicine. 450 B.C. Hippokrates mentiones in his oath that the treatment of patients with stones is to be left in the hands not of physicians, but of special men: "I will not cut persons labouring under the stone but I will leave this to be done by men who are practicineers of this work".

For thousands of years the history of urolithiasis has been the history of bladder stones. One of the intriguing mysteries of bladder stone is its frequency throughout medical history of the past and yet its rarity today. Indeed, a common cause of crying infants at night listed in the old textbooks was bladder stone - hardly the first thing a modern padiatrician would consider! Bladder stone has virtually disappeared from the children's hospitals today. This reduction in the actual number of bladder stones in Norfolk and Norwich Hospital is very striking and shows between 1871 and 1947 the disappearance of children's bladder stones by 1943 and a reduction in the number of adult bladder stones by over 80 percent.

Epidemiological studies are best done in geographically isolated regions without fluctuation of the population. The significant change in the pattern of urolithiasis is demonstrated here with the example of Sicily. Like in Norfolk in the 1940s the bladder stones disappeared almost completely and simultaneously the stones in the upper urinary tract appeared (Fig. 2).

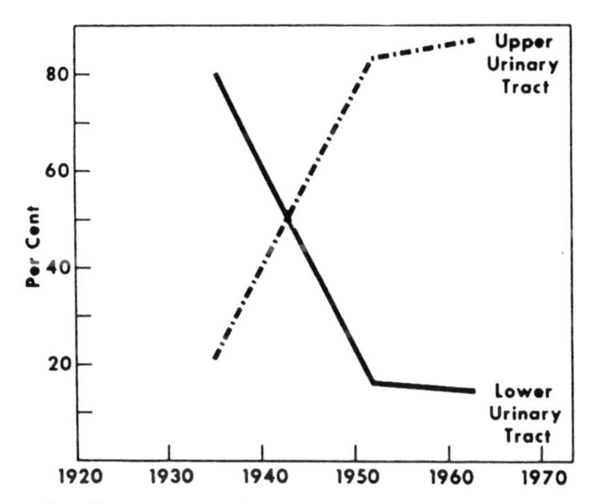

Reversal of pattern in anatomical distribution of urinary stones in Palermo, Sicily. (M. Pavone Macaluso, et al)

Fig. 2

On the other hand the history of urolithiasis has been for thousands of years the history of bladder stones.

The first exhaustive description of "cutting for the stone" has been given by Celsus. Celsus wrote on Lithotomy in the first century and indeed in the middle ages and later, the operation was often termed the Methodus Celsiana. Celsus himself advocated that the operation should only be performed on children between the ages of 9 and 14. The only suitable time of the year for the procedure was spring. The operation was allowed only if the child otherwise would have died. The technique then used is described in his "De Re Medicina" and was to remain almost unchanged for centuries.

The patient was held in the lap of a strong and intelligent person who steadied the patients by pressing his chest against his shoulder plates. If the child was large the single aide might be replaced by several persons with their legs tied together and additional assistance stood at their sides.

The operator stood or sat facing his victim and inserted the forefinger of the left hand, well dipped in oil, into the anus. The right hand was pressed on the suprapubic region, pushing on the bladder and thus forcing the stone into the grip of the left index finger within the rectum so as to produce a bulge in the perineum (Fig. 3).

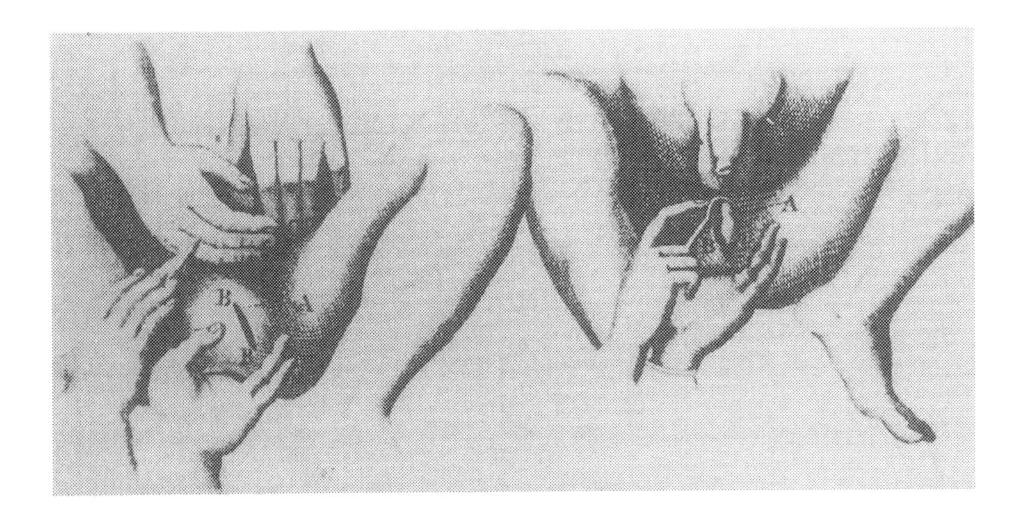

Fig. 3

A slightly curved transverse incision was made in front of the anus with the convexity directed towards the urethral bulb. The incision was carried deeply into the region of the bladder trigone and the stone was then pushed out by the finger in the rectum. It might be necessary at this stage to use a hook to dislodge the calculus. The wound was dressed with wool and warm oil.

This simple operation which involved the use of no staff in the urethra, no special instruments, merely a knife and perhaps a pair of forceps or a hook to help extract the stone, became known as the lesser operation or apparatus minor (Fig. 4).

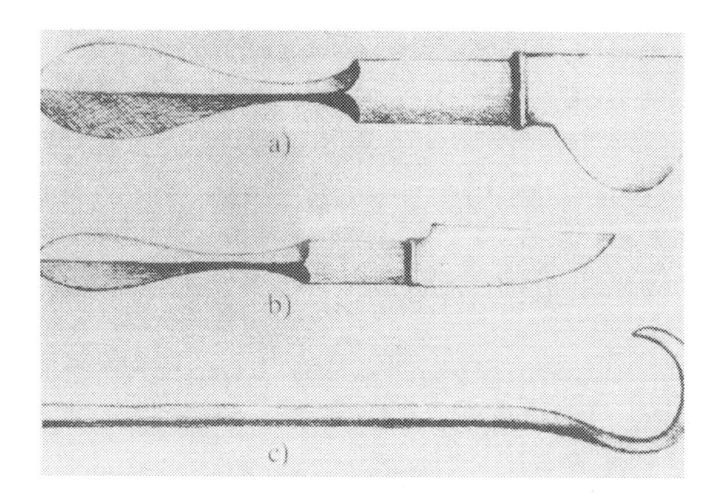

Fig. 4

Anatomically it involved opening the base of the bladder immediately above the prostate and it was for this reason that the operation was usually advised only for young boys, since in these subjects the prostate gland and seminal vesicals would be small.

The apparatus minor held the field until the early sixteenth century but it was still being described and carried out by surgeons in the mideighteenth century. This procedure, without any anesthetic help, had to be, like Celsus wrote in his original latin work: "Tuto, celeriter et jucunde."

This is very understandable. The problems appeared post-operatively. 50 % of the patients died immediately as a consequence of haemorrhage or infection. In cases with a better result urinary fistulas, fecal incontinence and impotency remained. Based on these figures the attitude of the church prevailing during the middle ages is understandable: According to the principle "ecclesia abhorret a sanguine" the operation was forbidden.

In 1520 a new technique of perineal lithotomy was introduced by Marianus Sanctus in his golden booklet "Libellus Aureus". Because of the additional instruments to be employed it was also termed the greater operation or the apparatus major. The plan of the procedure was to pass a grooved staff into the bladder and subsequently to cut down upon this instrument so that yet another name for the operation was "cutting on the staff". A vertical incision was made in the midline onto the groove into the staff to open the membranous urethra. A gorget was passed through the incision and this was followed by a series of instruments to dilate the wound, thus tearing through the prostate and bladder-neck.

The old surgical writers describe in depth the care and pre-causions in binding and holding the patient in the lithotomy position. "The patient shall be placed upon a firm table or bench with a cloth many times doubled under his buttocks, and a pillow under his loynes and back, so that he may lie halfe upright with his thighs lifted up and his legs and heels drawn back by to his buttocks. The patient, thus bound, it is fit you have 4 strong men at hand; that is 2 to hold his arms, and another 2 you may so firmely and straightly hold the knee with one hand and the foot with the other that he may neither move his limmes nor stirre his buttocks but be forced to keep the same posture with his whole body." It seems incredible to us in these days of smooth and potent anesthesia that anyone could possibly submit himself willingly to such torture. It was indeed only the terrible and protracted agonies produced by the stone that gave men sufficient courage to place themselves under the lithotomists cruel instruments (Fig. 5).

Fig. 5

The next step in the history of lithotomy is perhaps the strangest. It concerns the development of the lateral approach through the bladder by an unqualified Frenchman of humble origin who is unique in that his name has been perpetuated to this day in a well known nursery rime. Jacques Beaulieu was born in 1651 as the son of humble peasants in Burgundy. In 1690 he changed his name to Frere Jacques, adopted the habit of a monk, although he never trained for the church and he in turn became an itinerant lithotomist (Fig. 6).

Fig. 6

In 1697 at the age of 46 he arrived in Paris and applied for permission to cut for the stone. The surgeons at the Hotel Dieu ordered that he first demonstrate his skill on a cadaver in whom a stone had been introduced into the bladder via the abdomen.

He passed a solid grooveless metal staff into the bladder, then incised the perineum two fingers medial to the tuber ischii, carrying the cut forward from the side of the anus. The stone was felt by a finger in the wound, a dilator was passed into the bladder and the stone removed by forceps. In spite of the satisfactory operation the board refused to grant a license. This was probably because Frere Jacques paid no attention to the ritual of pre-operative bleeding or purging, using no adstringents but stated that he relied instead on God to heal the wound. Lateral lithotomy, even in this cruel form, was safer then the midline procedure in adults, since it gave wider access with less tissued trauma. Moreover, the pre- and postoperative treatment used by others at this time, probably did more harm then complete conservatism.

Frere Jacques, his license refused, travelled to Fontainbleau, where the court was in residence. Here he was allowed to operate on a shoemaker with a stone in the bladder. Cure was obtained within 3 weeks and Louis XIV was so impressed that he gave instructions that Frere Jacques be given the King's license to practice. A short period of success followed with its inevitable popularity, so that a guard of soldiers was required to keep the mass of spectators who crowded round in some sort of order. However, a series of disaster then befell the surgeon:

From April to July 1698 we carried out 42 lithotomies in the Hotel Dieu and 18 at the Charite. Of these 60 patients, 25 died, 13 were cured and the rest remained in hospital with incontinence, fistulae or other complications. No less then 7 died in 1 day at the Charite so that Frere Jacques was actually driven to accusing the monks of poisoning his patients. Postmortem, however, revealed no evidence of poison but did demonstrate bladders cut through in many places, the rectum injured, the urethra cut off from the bladder base, the vagina lacerated, and major arteries divided.

Another crisis occured in 1703, when he was consulted by the Marchall de Lorch, a cautious man, who first watched Frere Jacques operate upon 22 poor patients, all of whom survived, before submitting himself to lithotomy. All was in vain, since the Marchall succumbed after surgery.

Yet once again, Frere Jacques had to take to the road and finally arrived in Aachen. Aachen ist the Western most city in Germany and is the place where in the year of 800 Charles Magnus was crowned. Within 4 weeks Frere Jacques operated on 200 patients. The mortality was only 3 percent. The reason for that was that he operated in Aachen in clean homes and not in infected hospitals. Frere Jacques, who cutted on more than 4500 stones, is by sure the greates stonecutter of all times.

It seems incredible to us that the literature is replete with cases where patients experimented on themselves. One introduced a long nail into his bladder, impinged the end upon the calculus and struck hard with a black smith's hammer to split the stone. In the collection of instruments at the Royal College of Surgeons is the curved metal sound with a roughened edge which General Martin of Lucknow claimed, in 1783, to have gradually disintegrated the stone in his bladder by 9 months of steady work.

The most heroic case is the one of Jan de Dood from the Netherlands who is shown here. He already had two failed attempts of lithotomists to get rid of his stone. He decided like many others to operate on himself. His wife was sent to the fish market, only his brother was present when he removed his bladder stone (Fig. 7).

Fig. 7

Using this knife by himself he applied and reinventend the method of Celsus. The stone as well as the knife are now in the Museum of Medical History in Leiden, Netherlands. The patient survived.

In the year of 1708 the area of stone cutting ended. The general surgeons took over the business and there was no longer a reason to keep cutting on the stone as secret. Francois Tollet published the experience of all stone cutters in his classical work: "Traite de la Lithotomie".

Throughout the centuries, patient teased by the agony of bladder stone and surgeons dissatisfied with the difficulties and dangers of cutting for the stone dreamed of some means of removing the calculas through the urethra. Jean Civiale had already commenced a series of experiments in 1817 when he was still only a second year medical student in the University of Paris, in an effort to ascertain whether it was possible to crush stones in the bladder without injuring its walls. To him falls the honor of performing the first successful lithotrity on 13 January, 1824 at the Necken Hospital in Paris.

Civiale's instrument, the trilabe was introduced blindly into the bladder, there he made numerous separate perforations into the stone (Fig. 8).

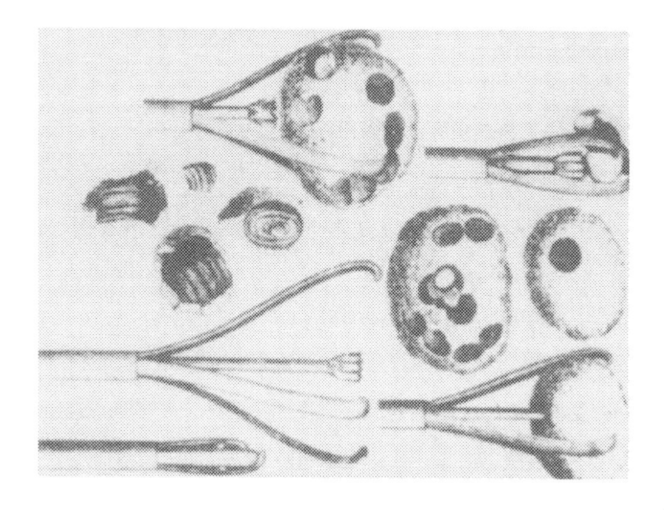

Fig. 8

The drill of Civiale's instrument was rotated by a violin bow.

This is an original drawing of Civiale from 1847 showing a patient and the instrument in place. We should not forget that the major advantages of modern surgery - anesthesia and asepsis were unknown (Fig. 9).

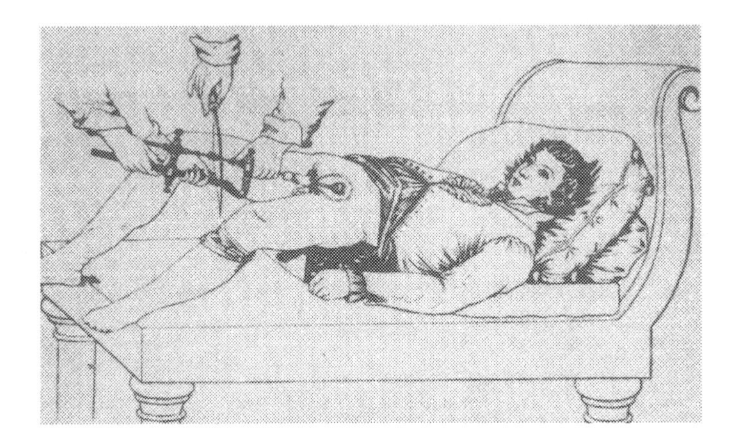

Fig. 9

There is no doubt that at this particular period in history which span the pre-anesthetic, pre-listerian days over into the pioneer period of anesthesia and antisepsis the lithotomy was an enormous advance in the treatment of bladder stones.

A major disadvantage of all transurethral lithotomies was that they had to be done blindly. October 2, 1877 was the day when Nitze presented his instrument the first time before the Royal Medical Society in Dresden, now East Germany. He impressed the society but the handling of the cystoscope seemed to be too difficult. Nitze himself tells the circumstances in which he solved the problems and this again shows that one of the important qualities in research is observation. One day, in the hospital in Dresden, while examining the objective of a microscope to see whether it was clear he looked through the lense at a neighboring church and saw the upside down image of the church. Immediately, the idea occured to him that he could easily obtain an enlargement of the visual field with a system of lenses. With the help of the instrument maker he incorporated these ideas in making the first cystoscope.

It consisted of a metallic catheter with one end bent like a hockey stick (Fig. 10).

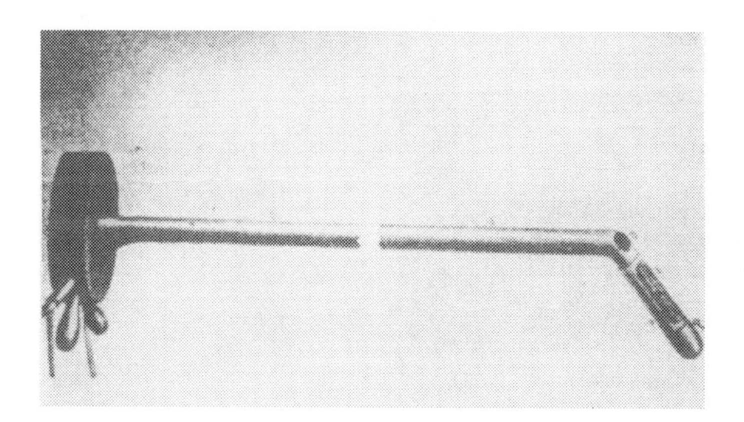

Fig. 10

Its caliber corresponded to 21 French. An incandescent platinum wire was located near the tip and illuminated the bladder. The instrument was complicated by the fact that it was necessary to cool the wire by circulating water in order to avoid burning of the bladder. The urologists needed a porter to carry all the equipment necessary for cystoscopy.

In 1900, Nitze became Professor at the University of Berlin and Berlin became a Center of Urology.

Another major breakthrough in bladder stone treatment was the advent of ultrasonic lithotripsy. Since 1968 technicians and physicians of the Technical University in Aachen, used ultrasound experimentally and clinically in the treatment of bladder stones.

The unit is composed of 3 elements: A cystoscope and an angulated optical system, a pieco electric ultrasonic transducer and an interchangable ultrasonic probe. In a resting position the drilling probe is completely withdrawn in the shaft. During the operation the probe is at once under visual control from the shaft towards the stone (Fig. 11).

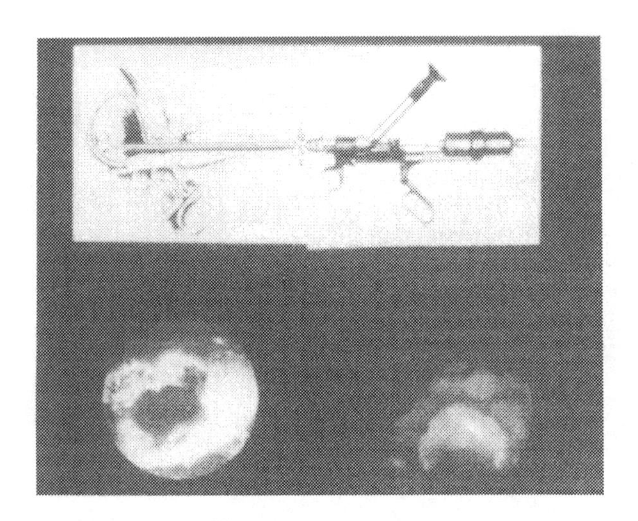

Fig. 11

The desintegration of the stone is observed through the cystoscope telescope. The ultrasonic generator is triggered by a footswitch. The hollow ultrasonic probe is connected to a sucction pump via a hose to remove the minute stone fragments. Irrigation is performed through the cystoscope. Meanwhile, this instrument is not only the basis of bladder stone destruction but also of all types of percutaneous surgery.

The bladder stone from 4800 B.C. has been the world's oldest calculus. It sounds like a bad choke and is sad to relate. This unique specimen and oldest stone at all was already destroyed by a shockwave. However, this was the uncontrolled shockwave emerging from a bomb when the Royal College of Surgeon in London was destroyed in 1941. I believe, nothing could demonstrate better the benefits of the peaceful use of shockwaves we have achieved now.

A Survey of the Action of Lasers on Stones

G. M. Watson

The Middlesex Hospital, Department of Urology,
Mortimer Street, London WIN 8AA, United Kingdom

Summary

Continuous wave lasers vaporise calculi relatively ineffectively via fibres since temperatures of > 3000° C are required. A train of pulses at just above the fragmentation threshold is preferred to single giant pulses which tend to propel the stone violently. Pulse durations in the range of 10 nanoseconds are difficult to transmit through fibres regardless of wavelength. The bare fibre cannot be used close to a stone because it tends to disrupt. Microsecond pulses cause the most elimination of material per pulse and are easy to transmit via fibres. The bare fibre can be held in contact with the stone surface. The optimum parameters for stone fragmentation in clinical practice are using microsecond pulses via fine quartz fibres touching the stone surface. A plasma is formed just beyond the tip of the fibre only when there is sufficient absorption. By selecting the wavelength at close to 500 nm there is maximal differential absorption so that a plasma forms readily at the stone surface but not at the ureteric wall.

Introduction

There are already techniques which exist for the treatment of stones in the ureter. The least invasive modality is to use extracorporeal shock wave lithotripsy (ESWL). This modality makes use of supersonic waves which are focussed through the body wall onto the stone and thus is the least invasive of all therapies. It is less effective for stones in the ureter than for stones in the renal pelvis or calyces. This is because when a stone is impacted in the ureter the walls of the ureter damp down the shock waves and secondly because of the problems of imaging the ureteric stone. ESWL can actually cause ureteric obstruction as a result of impaction of debris. Endoscopic modalities are likely to be required for the foreseeable future.

Endoscopy in the ureter by insertion of ureteroscopes has become a widespread skill. Initially the only technique for dealing with a calculus was to trap it in a wire basket under vision and then manoeuvre it back out of the ureter. Stones which were tightly impacted in the ureter could not be extracted. The development of modalities for stone fragmentation which can be used via ureteroscopes has made ureteroscopy far more useful. The first modality was ultrasound. The piezo-electric crystal drives a hollow, rigid wand which is inserted through the instrument channel of the endoscope. The wand merely vibrates and has a drilling action on the stone. The device requires a significant flow of irrigation throughout its use to prevent heating to levels which can damage tissue. Firm contact is required with the stone which can be pushed back up the ureter. 13.5 and 11.5 F endoscopes are required to accommodate the relatively large wands. These larger endoscopes are more difficult to insert and are more traumatic than smaller endoscopes (Watson et al. 1987). The electrohydraulic probes are miniaturised spark plugs which set up a local shock wave when discharged underwater. 5F electrohydraulic probes can be used via 9.5 and 11.5 F ureteroscopes. The electrohydraulic action is a controlled explosion. If the probe is discharged too close to the ureteric side wall then it causes a significant perforation.

Any new technique must offer a significant advance over the modalities described above. This may be by advantages in the efficiency of fragmentation or by its impact on the endoscopes. It may also expand the sphere of action of the endoscopic fragmentation to the biliary system.

Materials and Methods

Human urinary and biliary calculi were taken to laser centres chosen so that as wide a variety of laser parameters as possible could be tested. Any laser action will be affected by the method of delivery of the laser energy. Therefore wherever possible the lasers were transmitted through fibres.

[1] Continuous Wave Lasers:

a) Neodymium YAG

The laser tested was capable of an output of 70 W via a 600 micron fibre. The emergent beam diverged through approximately 16°. The action on stones composed of calcium and magnesium compounds in air was to cause vaporisation leaving behind a white ash residue which formed a barrier to any further laser action. If the fibre was brought close to the stone surface then it was damaged within a few seconds as a result of overheating of the quartz. In water there was a very slow action only with the fibre close to the stone. Cystine, urate and the majority of biliary calculi were vaporised much more efficiently without any ash formation. In air the stones became hot. In water the action was damped considerably and required close proximity of the fibre to the stone. This laser is ineffective for the majority of calculi and is unsafe to use in the ureter because of the devastating effect on the tissue associated with these high energies in a confined area.

b) Argon

The argon laser was capable of delivering 15 W via a 600 micron fibre. This laser was more effective than a neodymium YAG laser at 15 W and was equivalent to that laser at approximately 25 W on pale coloured stones. Again this laser is not suitable for use on calculi.

c) Carbon dioxide

The carbon dioxide laser had a power of 15 W. When focussed at a power of 10 W with a spot diameter of 0.5 mm it drilled through calculi. Again the action was solely thermal. The stone in the immediate focus of the beam was vaporised. Surrounding the crater was a zone of carbonisation and white ash. In contrast to the other wavelengths the carbon dioxide laser could vaporise the ash given a sufficient energy density. Even with the improved absorption at this wavelength the calculi became hot to handle within 50 seconds application. cholesterol gallstones could be ignited.

[2] Q Switched Lasers

a) Ruby

This laser could deliver single giant pulses of up to 7 J in a 20 nanosecond duration. When this laser was focussed to a 0.5 mm spot diameter on calculi in air or water it caused a violent propulsion with only occasional splitting of the stone. Gallstones and struvite stones tended to be split while more dense calculi were resistant.

b) Neodymium YAG

This laser delivered pulses of up to 800 mJ in a 15 nanosecond pulse duration a repetition rate of 10 Hz. When focussed to a spot diameter of 0.5 mm the laser had a threshold for fragmentation of 150 mJ. The action of a single pulse on a stone was very small with negligible propulsion. However when repeated at a repetition rate of 10 Hz there was a steady ablative effect. Each pulse caused a plasma at the stone surface. 2000 pulses delivered to a gallstone resulted in a significant crater but 20 000 pulses delivered to a dense calcium oxalate (principally monohydrate) stone resulted in a negligible effect. It was possible to transmit this laser through a 600 micron fibre with considerable difficulty. Whenever the distal end of the fibre was brought close to the stone surface to test its action on calculi the fibre tip disrupted before any effect on the stone was observed.

c) Frequency doubled neodymium YAG

This laser delivered pulses of up to 400 mJ in a pulse duration of 15 nanoseconds at a pulse repetition rate of 10 Hz. When focussed to a spot diameter of 0.5 mm this laser had an identical threshold and action to the laser at fundamental (1064 nm). There was a similar plasma formation and a similar slow ablation of the calculus.

d) Carbon dioxide

This laser was capable of emitting pulses of up to 1 J in a pulse duration of 10 nanoseconds at a repetition rate of 30 Hz. At these energies the laser had an effect in air only on all calculi with a beam 5 mm in diameter. There was a spray of very fine particles released from the stone surface. After only one minute of application the stones became hot unless the pulse repetition rate was kept to 10 Hz or less.

Since Q switched lasers were difficult to transmit via fibres the two possible solutions are to either transmit lower pulse energies at a higher repetition rate or to turn to lasers with longer pulse durations [3]. The copper vapour laser provides a combination of very high repetition rate with low pulse energy although with short duration pulses. The excimer laser wavelengths are absorbed very efficiently and might therefore be used at a lower pulse energy.

e) Copper vapour laser

This laser delivered pulses of up to 5 mJ with a pulse duration of 20 nanoseconds and a repetition rate of 4 KHz. The output could be coupled into a fibre 600 microns in diameter with 70% efficiency. The action of the laser on calculi was to similar to that of a continuous wave laser with vaporisation, melting and charring of the stone. There was no fragmentation.

f) Excimer lasers

The following excimer laser mixes were tested, argon fluoride, krypton fluoride, xenon chloride and xenon fluoride. The argon fluoride laser was focussed onto a variety of calculi in air. With a spot size of 0.2 mm^2 the threshold for ablation was 12 mJ (6.1 J/cm^2) but with a spot size of 1.25 mm^2 the threshold was 25 mJ (2.0 J/cm^2). The laser had no effect on calculi in water. When used at a repetition rate of over 30 Hz there was some carbonisation due to heating of the surrounding stone. The krypton fluoride mix focussed in air to a spot size of 1.0 mm^2 had a threshold of 10 mJ (1.0 J/cm^2). It was less effective underwater. The xenon chloride gas mix when focussed in air to a spot size of 1.0 mm^2 had a threshold of 12 mJ (1.2 J/cm^2). In water the threshold for the same spot size was 20 mJ (2.0 J/cm^2). The xenon fluoride gas mix had an identical threshold to the xenon chloride with a 1.0 mm^2 spot size. However when the laser was used with a spot size of 4 mm^2 the threshold was 17.5 mJ (only 0.44 J/cm^2) in air and 30 mJ (only 0.75 J/cm^2) in water. The xenon fluoride laser was coupled into a 600 micron quartz fibre with transmission of up to 30 mJ pulses for a period of 1 minute. There was a more efficient action on the stone in water than could be achieved with the focussed beam. The fibre had to be maintained 5 mm from the stone. **In the region** of 10 micrograms were ablated per pulse. There was little propulsion. The particles were extremely fine and most of the matter was released as vapour. The ablative action was more equivalent to drilling than to fragmenting the stone.

A giant pulsed krypton fluoride laser was used at energies of several joules on a stone. There was a giant plasma formation with considerable propulsion.

[3] Lasers with intermediate pulse durations

a) 1 microsecond pulsed carbon dioxide

This laser was capable of delivering single pulses of 1 J. It liberated a spray of fine particles on all calculi in air only. There was not a significant difference in comparison to the 10 nanosecond pulsed laser.

b) 100 microsecond neodymium YAG

This laser emits pulses of up to 1.5 J with a maximum average power output of 30 W. It proved possible to transmit this laser through a 600 micron fibre with 90% efficiency but with only 50% efficiency through a 400 micron fibre. When used in air the laser had a thermal effect which was indistinguishable from the 70 W continuous wave neodymium YAG laser. However when the laser was used underwater then the fibre could be held touching the stone without the same thermal damage seen with the continuous wave laser. There was for the first time a fragmentation effect on certain stones. The threshold for this fragmentation was dependent on the pulse energy rather than the average power output. The threshold using a 600 micron fibre was 1 J for struvite calculi. There was negligible action even at 1.5 J on calcium oxalate. The threshold for dark biliary calculi and for charred areas on calculi was approximately 500 mJ. Thus it proved possible to carbonise the surface of certain calculi in air and then use the laser fibre in contact with the charred portion underwater. Several cycles of applying the laser in air then water could break a typical ureteric calculus of calcium oxalate (dihydrate) into 5 fragments. When the 400 micron fibre was used there was a drop in threshold to approximately 70% of that using the 600 micron fibre. However the fibre was damaged more rapidly. The action of the laser when used directly on tissue was to cause significant and penetrating coagulation. Alternative laser systems which could fragment calculi without prior carbonisation (therefore using a wavelength which is better absorbed by the pale coloured calculi) were sought.

c) Long pulsed ruby

This laser produced pulses of up to 1.5 J in a pulse duration that varied from 10 microseconds up to 800 microseconds as the pulse energy was increased. The pulse repetition rate was less than 1 Hz. The output from the laser was coupled with a 600 micron fibre with only 50% efficiency due to the relatively poor beam quality. There was negligible action on pale calculi even with the fibre in contact with the stone. The fibre penetrated dark cholesterol gallstones without fragmenting them.

Spectroscopic analysis of urinary calculi was performed using a Beckman spectrophotometer with an integrating sphere. This showed that urinary and biliary calculi absorb powerfully in the ultraviolet and absorb minimally in the 1000 nm region. There was a gradual drop in absorption between these points in the case of urinary calculi. Biliary calculi had a secondary peak of absorption at 450 nm. There were therefore a number of reasons for assessing the pulsed dye lasers. The pulsed dye lasers can be used to test a wide range of pulse durations and a range of wavelengths spanning the visible spectrum.

d) **Pulsed dye lasers**

The pulsed dye lasers allowed a parametric study on fragmentation to be conducted. The wavelengths of the laser output were varied from 445 to 577 nm and the pulse duration from 1 microsecond to 360 microsecond. The output from the laser was coupled into quartz fibres with transmission efficiency of 90% with 1 mm and 600 micron fibres and 70% efficiency for 200 fibres. At pulse energies above threshold and with the fibre in contact with the calculi underwater fragmentation occurred without carbonisation. There was no action without confinement by water. As the pulse energy was increased so the action altered from the creation of a small pit in the stone surface to crater formation to break up of the entire stone. There was a prominent "click" sound associated with the formation of a bright white light indicating plasma formation.

The fragmentation threshold was measured as each parameter was varied in turn. The threshold was significantly higher for all stones at 577 nm than at 504 nm and at 504 nm than at 445 nm with all other parameters held constant. The threshold was significantly higher with every stone for each lengthening of the pulse duration of the laser from 1 to 10 to 120 to 300 to 360 microseconds with all other parameters held constant.This effect was not proportional to the power however. A 300 fold alteration in power produced only a 10 fold alteration in the threshold. The fibre diameter was approximately the diameter of the spot on the stone because the fibre was held in contact with the stone. The threshold was significantly higher for every stone as the fibre diameter was increased from 200 to 400 to 600 to 1000 microns. The effect of the fibre diameter was proportional to the energy density.

The parameters for stone fragmentation at the minimum energy are a wavelength of 445 nm, a pulse duration of 1 microsecond and a fibre of 200 microns. The threshold for a typical calcium oxalate dihydrate calculus is 15 mJ. The optimum parameters for fragmentation in the clinical setting is to make use of the wavelengths in the green region where there is maximal differential absorption between the stone and the ureter. Using 504 nm and the 1 microsecond pulse duration via a 200 micron fibre the threshold is 20 mJ. Further testing of calculi has shown that calcium oxalate monohydrate stones fragment more efficiently at pulse energies above threshold using a 320 micron fibre instead of a 200 micron fibre. This is an idiosyncratic result and no explanation has yet been found. Oxalate stones consisting of almost 100% monohydrate are extremely resilient to fragmentation with any shock wave modality. The ability to fragment these and most cystine stones is an advantage.

Conclusions

Certain trends were apparent from this survey:

1 Continuous wave lasers have a thermal action alone.

2 Pulses in the region of 10 nanoseconds are difficult to transmit via fibres. There is no possibility of using the bare fibre in contact with the calculus at these pulse durations.

3 Single giant pulses tend to cause excessive propulsion of the stone. It is preferable to utilise pulses at an energy just above threshold at a pulse repetition rate.

4 At high repetition rates (4 KHz in the case of the copper vapour laser, 30 Hz in the case of the 10 nanosecond carbon dioxide laser and 30 Hz in the case of the argon fluoride laser) pulsed lasers are more likely to have a thermal effect.

5 Fragmentation can be achieved via fibres using pulses of between 10 nanoseconds and 100 microseconds. At longer pulse durations absorption is a prerequisite for fragmentation.

6 Although a plasma is probably the basic mechanism responsible for fragmentation there are certain distinctions which can be made. Pulse durations of 1 microsecond or longer are only effective when confined underwater. 10 nanosecond pulses cause plasmas which appear to be equally effective in air and water. The plasma resulting from a 1 microsecond pulse removes more material per pulse than that of other pulse durations.

7 Only the microsecond pulses have a sufficiently destructive action to fragment calcium oxalate monohydrate stones.

Discussion

There are a number of possible combinations of laser parameters which can be used to fragment calculi. The final choice of laser parameters is made according to clinical expediency. The system must fragment calculi efficiently. There must be no risk of injury to surrounding tissue as a result of fragmenting the calculus or of stray pulses on the ureter or common bile duct. In order to have advantages over existing modalities of stone fragmentation the laser system must have potential implications on the ureteroscope required to deliver the laser fibre to the stone.

One possible advantage of a laser system is in the miniaturisation that the quartz fibre offers. The 200 and 320 micron fibre of the pulsed dye laser allows some significant miniaturisation of ureteroscopes. Ureteroscopes have a common channel for irrigation and for passage of instruments. The electrohydraulic and ultrasound probes reduce the flow of irrigant significantly when passed. The fibre of the pulsed dye laser however can be passed through even the smaller ureteroscopes without significantly reducing the flow of irrigation. The smaller the ureteroscope the lower the risk of damaging the ureter (Watson 1987). The flexibility of the fine fibre allows it to be used via the instrument channel of flexible ureteroscopes which makes it particularly useful for antegrade ureteroscopy. If a laser system was used requiring a larger diameter fibre (as for example with a Q switched laser system) then this clinical advantage would be lost.

Absorption is more significant for fragmentation when using microsecond pulses and longer. If a laser system could be used without ureteroscopy then it would be a significant advantage since ureteroscopy is difficult to perform and potentially damaging to the ureter. The plasma from the pulsed dye laser system is formed only when there is sufficient absorption at the stone surface. This is one of the causes of the reduced action of the laser on tissue rather than stone. This raises the possibility of using the laser without endoscopy and using the detection of the plasma extracorporeally by monitoring the plasma. This might be by acoustic detection or by a second fibre to record the plasma emission.

Thus although there are many possible parameters for fragmenting calculi only the pulsed dye laser can fragment calcium oxalate monohydrate stones, can be transmitted via very fine fibres and can be used with a bare fibre. Only the pulsed dye laser can be used in the green where there is differential absorption by the stone over tissue. Consequently only the pulsed dye laser forms a plasma on the stone and not on the normal ureter.

References

Laser fragmentation of renal calculi. G.M.Watson, J.E.A.Wickham, T.N.Mills, S.G.Bown, C.P.Swain, P.R.Salmon. British Journal of Urology (1983) 55: 613-616.

Laser fragmentation of urinary calculi. G.M.Watson. In "Lasers in Urologic Surgery" 1985. Ed. J.A.Smith. Year Book Medical: Chicago.

An overview of the action of lasers on calculi - Laboratory and Clinical studies. G.M.Watson, J.E.A.Wickham. Fortschritte der Urologie und Nephrologie (1986) 25. 359-368. Ed Vahlensieck, Gasser.Steinkopff Verlag, Damstadt.

The pulsed dye laser for fragmenting urinary calculi. G.M.Watson, S.Murray, S.P.Dretler, J.A.Parrish. Journal of Urology (1987) 138: 195-198.

Laser Induced Breakdown Spectroscopy (LIBS) of Kidney Stones

W. Meyer[1], *R. Engelhardt*[1], *and P. Hering*[2]

[1]Medizinisches Laserzentrum Lübeck, Peter-Monnik-Weg 9, D-2400 Lübeck, Fed. Rep. of Germany
[2]Max-Planck-Institut für Quantenoptik, Postfach 1513, D-8046 Garching, Fed. Rep. of Germany

Laser induced shock wave lithotripsy (LISL) using a flashlamp pumped dye laser produces a plasma on the stone surface which leads to a bright flash of intense light. The high temperature of the plasma vaporizes the stone material leading to electronically excited atoms and ions.

Spectrally dispersed light displays a broad continuum superimposed by atomic emission and absorption lines. Temporally resolved plasma spectra exhibit spectrochemical information about the composition of the stone material.

We use three 200 µm core diameter quartz fibers, one to transmit 40 to 80 mJ pulses from a flashlamp pumped dye laser (590 nm, 2 µs pulsewidth) to the stone surface, the other two to guide the plasma emission back from the stone to the entrance slit of a 25 cm spectrometer and to a photodetector. A gateable, intensified optical multichannel analyser is used to detect plasma spectra yielding an excellent signal with only a single laser shot.

In vitro we demonstrated that the obtained spectra are stone specific and can be used for stone analysis simultanously to fragmentation.

1. Introduction

We investigated the possibility of a spectral analysis of kidney stones concomitant to the laserinduced shock wave lithotripsy (LISL) with a flash pumped dye laser.

The physical mechanism of shock wave induction by laser is that a laserpulse with a high energy density focused on a target induces a high electromagnetic field. Due to the multiphoton ionization of atoms and molecules of the target material free electrons are produced. The increasing free electron density as the result of an avalanche ionization leads to a plasma formation with temperatures greater than 10000 degrees and is accompanied with a bright flash [1]. The expansion of the plasma induces a spherical shock wave leading to mechanical ruptures on the target (Fig. 1).

Physical principles of laser-induced breakdown
● Short laser pulse focused at target (10^{-6}–10^{-8}s)
↓
● High power density ~10^{12}W/cm^2
↓
● High electric field ~10^6V/cm
↓
● Plasma formation ~10^{21} free electrons/cm^3 >20 000°C
↓
● Shock wave >400 bar
↓
● Localized mechanical rupture for small radius

Fig. 1

For the lithotripsy with a flash pumped dye laser the energy is coupled into a quartz fiber with a core diameter of 200 µm. The mean power density is 70 MW/cm^2. A direct contact of the fiber end with the stone is needed to produce a dielectrical breakdown. This is proved with the mechanical ruptures on the surface of the stone and by the bright flash.

The spectrum of the plasma flash consists of two components, a continuum superimposed by emission and absorption lines of atoms and ions of the vaporized and dissociated stone material. Teng describes the temporal progress of a plasma flash beginning with a continuum so high that line spectra are not visible. Then the continuum decreases and the line spectrum rises out of it [2]. This means by analysing the line spectrum at an optimal time there should be a possibility to get an information about the target material.

2. Results

The experimental configuration to get plasma spectra is shown in Fig. 2. The laser energy was coupled with a 200 µm fiber onto the stone positioned in water. A part of the reemitted light was guided back with two other fibers. One fiber was connected to a photo detector to look for the temporal progress of the plasma flash. The photodetector signal was recorded with a digital oscilloscope. The second fiber was connected to a gatable spectrometer. The delay circuit for gating the spectrometer was connected to the trigger output of the laser to determine the time of spectrum analysing. The gate pulse was also recorded on the digital oscilloscope. The laser

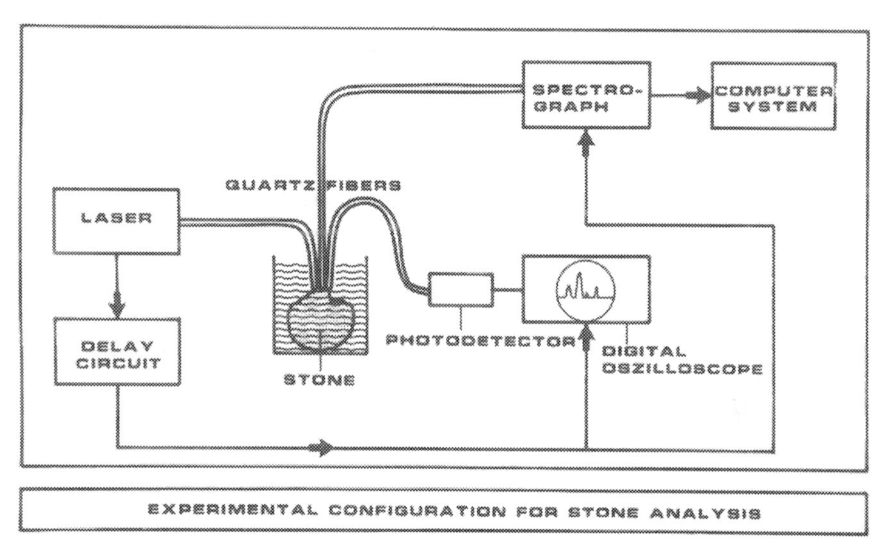

Fig. 2

we used is a flashlamp pumped dye laser working with Rhodamine 6G. The wave length is about 590 nm. We worked with pulse energies of 45 mJ and pulse lengths of about 2 µs. The spectrometer we used consists of a monochromator and a gatable optical multichannel analyser. A spectral range of 400 nm was imaged onto 1024 channels with the grating we used. The spectrometer was adjusted to detect the wavelengths between 180 and 580 nm.

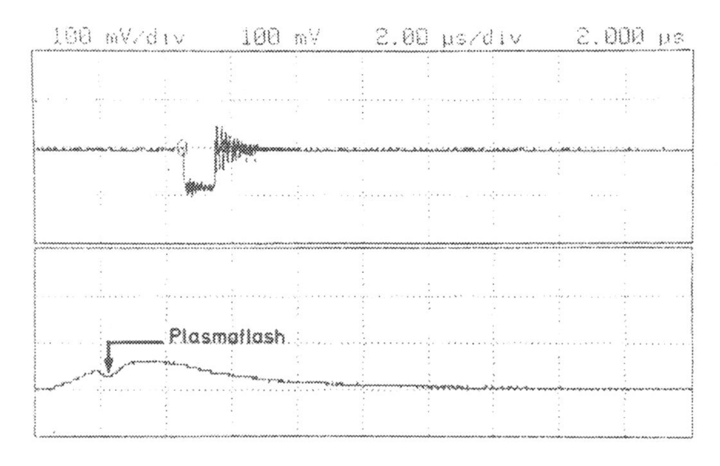

Fig. 3 Photodetectorsignal and spectrometer gating

As described before at the beginning of the plasma flash a broad continuum superimposed by small emission and absorption lines occurs in the spectrum. At the end the signal would be to small for detection. So we recorded first a temporal spectrum. Figure 3 shows the photodetector signal and the gate pulse of the spectrometer. The photodetector signal consists of laser light at the beginning of the laser pulse. After about 3 µs the light of the plasmaflash superimposes on the laserpulse. We recorded spectra at different times.

Figure 4 shows a spectrum of a calcium oxalate stone. It was recorded 2,5 µs after the beginning of the laserpulse. One can see a continuum with small absorption and emission lines.

Figure 5 shows a spectrum of the same stone as before 4 µs after the beginning of the laser pulse. The continuum has decreased and the emission and absorption lines of atoms and ions - most of them could identificated as sensitive lines of calcium atoms and ions - have increased. So we saw the exspected results of the temporal progress of the laser plasma.

In Fig. 6 a spectrum of an uric acid stone recorded 4 µs after the beginning of the laser pulse is shown. One can see the same emission and absorption lines as in the calcium oxalate stone. This means there must also be calcium in this stone.

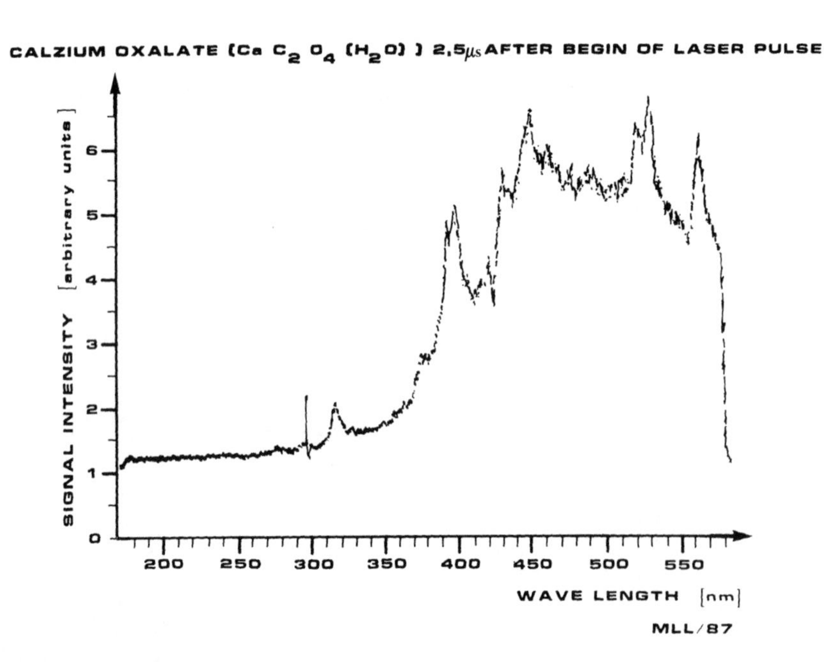

Fig. 4

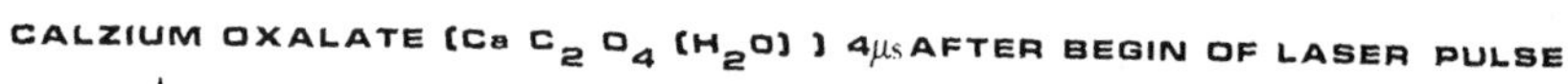
CALZIUM OXALATE (Ca C2 O4 (H2O)) 4μs AFTER BEGIN OF LASER PULSE

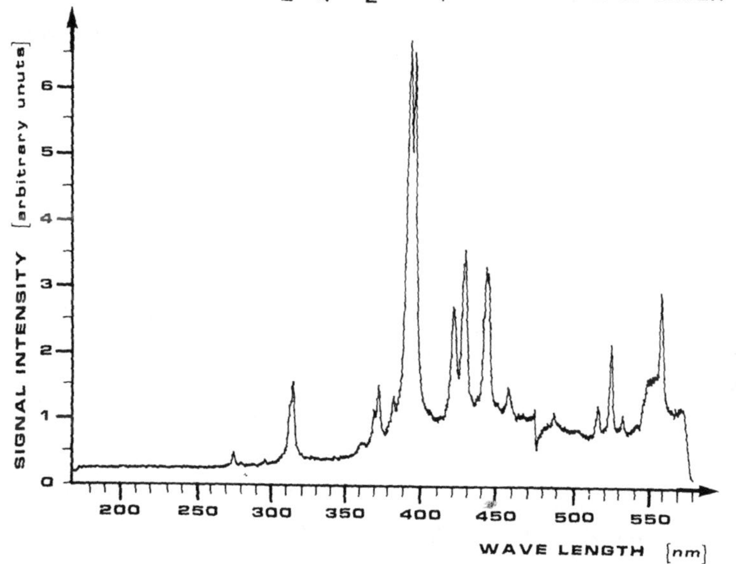
SIGNAL INTENSITY [arbitrary units]
0
1
2
3
4
5
6
200
250
300
350
400
450
500
550
WAVE LENGTH [nm]

Fig. 5

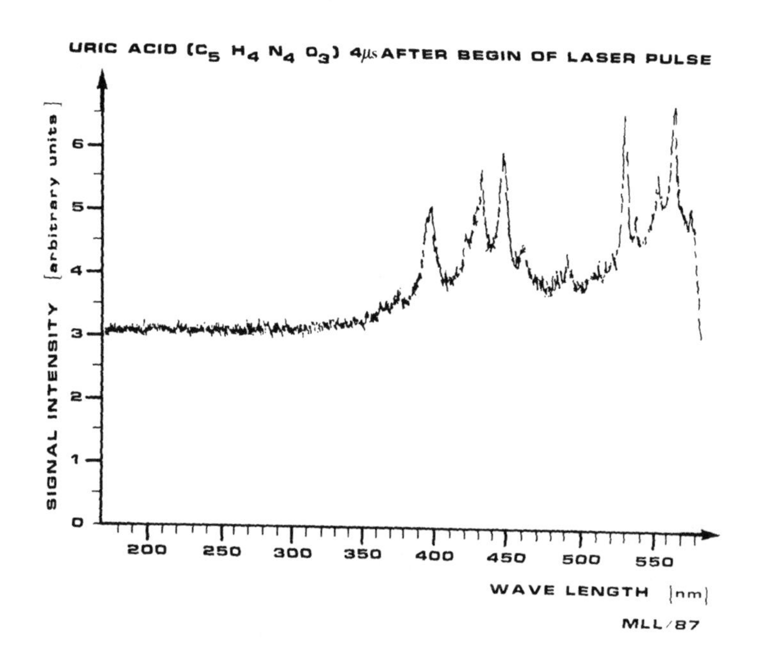
URIC ACID (C5 H4 N4 O3) 4μs AFTER BEGIN OF LASER PULSE
SIGNAL INTENSITY [arbitrary units]
0
1
2
3
4
5
6
200
250
300
350
400
450
500
550
WAVE LENGTH [nm]
MLL/87

Fig. 6

The similarity of spectra from stones of different composition shows that quantitativ analysis of stone material is extremly difficult. However, the relative peak intensities in stone spectra may be used to find a algorithm for stone analysis.

3. Conclusion

The possibility of spectral analysis of the plasma flash during the laser induced shock wave lithotripsy with a flashpumped dye laser was discribed. The temporal progress of the plasma flash was investigated to get a good resolved line spectrum. It could be shown that the detected spectra consist of a continuum superimposed by absorption and emission lines of atoms and ions. These results give a possibility for the calculation of the composition of a kidney stone.

4. References

1. Radziemski, L. J., Solarz, W. S., Paigner, J. A.
   "Laser spectroscopy and its applications"
   Marcel Dekker, Inc. New York and Basel

2. Teng, P., Nishioka, N. S., Rox Anderson, R., Deutsch, T. F.
   " Optical studies of pulsed laser fragmentation of biliary calculi," p. 73 - 78 Applied Physics B 42 (1987)

Identification of Body Concrements by Fast Time-Resolved Spectroscopy of Laser Induced Plasma

J. Helfmann[1], *H.-P. Berlien*[1], *T. Brodzinski*[1], *K. Dörschel*[1], *C. Scholz*[2], *and G. Müller*[1]

[1]Laser-Medizin-Zentrum GmbH Berlin, Krahmerstr. 6–10, D-1000 Berlin 45, Germany
[2]Freie Universität Berlin, Fachgebiet für Biomedizinische Technik, Schwerpunkt Lasermedizin, D-1000 Berlin, Germany

A method is presented to distinguish between different kinds of body concrements. The time resolved spectrum of a plasma generated by a laser pulse on the stone surface is recorded. These spectra are shown to be characteristic for the different kinds of body concrements. The influence of the laser wavelength, pulse energy and timing of the detection on the emission spectrum is discussed.

1 Experimental Set-Up and Method

Several human urinary and biliary calculi were positioned under water. The laser beam from a Nd-YAG laser (Spectra Physics DCR 3) using either a wavelength of 1064 nm or 532 nm and a pulse length of about 8 ns was focussed on the stone surface via a 80 mm focussing lens (figure 1). The focal area achieved in water was about 0.005 mm^2. Above a certain energy threshold a plasma is generated on the stone surface. The atomic, ionic and molecular emission lines provide information about the content of this plasma.

Via an optical fibre cable the emitted light is transmitted to an optical multichannel analyser (OMA) which is shown in figure 2. In the OMA a spectrum of the emitted light is analyzed in the selected region from 280 - 660 nm. The synchronisation of the OMA with the laser pulse is effected with a pretrigger signal from the laser and various time delays. Then the detector is activated for a period of about 150 ns, within which time the spectrum is obtained.

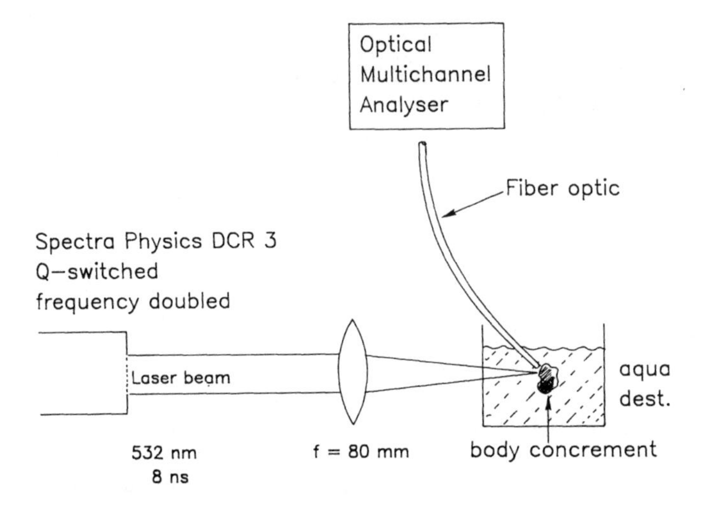

Fig. 1: Experimental set-up for time resolved spectroscopy of the laser induced plasma under water

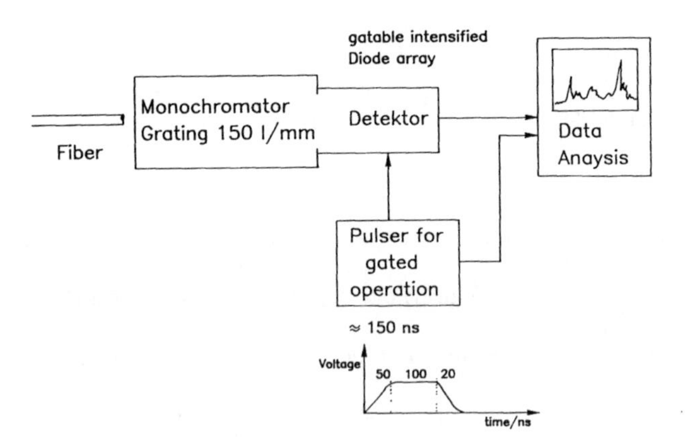

Fig. 2: The optical multichannel analyser (OMA III EG & G) consists of a monochromator-detector unit with a wavelength resolution better than 1 nm, a pulser which activates the detector for small periods of time and a computer for data analysis and timing

2 Experimental Results

In figure 3 five typical emission spectra from different body concrements are shown. Apart from some similar structures, they show differences in emission which could be used to identify the body concrements.

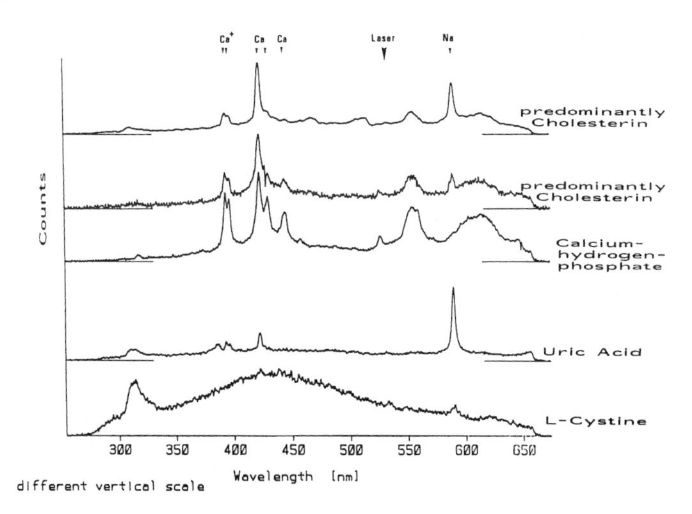

Fig. 3: Emission spectra of 5 different body concrements, whereby each spectrum is the sum of 10 single spectra produced by one laser pulse. Because of the different emission intensities, the vertical scales for the spectra differ from one another. On the right hand side of each spectrum the composition of the stone is denoted according to x-ray diffraction analysis

In order to understand these spectra better further investigations are necessary to establish the dependence of the spectra on the experimental parameters. These are: the wavelength of the excitation laser, the time delay between laser pulse and detection of the emitted light and the energy of the initial laser pulse.

3 How Do the Spectra Depend on the Excitation Wavelength ?

Both figure 4 and 5 show a comparison between the emission spectra of a body concrement obtained using two different excitation wavelengths - 532 nm and 1064 nm.

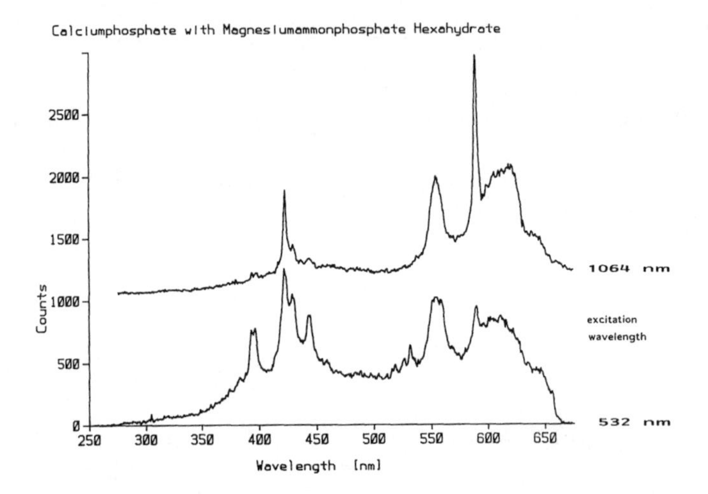

Fig. 4: Comparison of two emission spectra from the same body concrement but with a different excitation wavelength, the upper spectrum with 1064 nm, the lower one with 532 nm

Independent of the excitation wavelength, both spectra show the same typical structures. Even the Calcium-ion lines at 393 nm and 396 nm can be seen in the upper spectrum of figure 4 but only very weak compared to those in the spectrum obtained at the 532 nm excitation. This can also be seen for several other lines in the ultraviolet and blue part of the spectrum

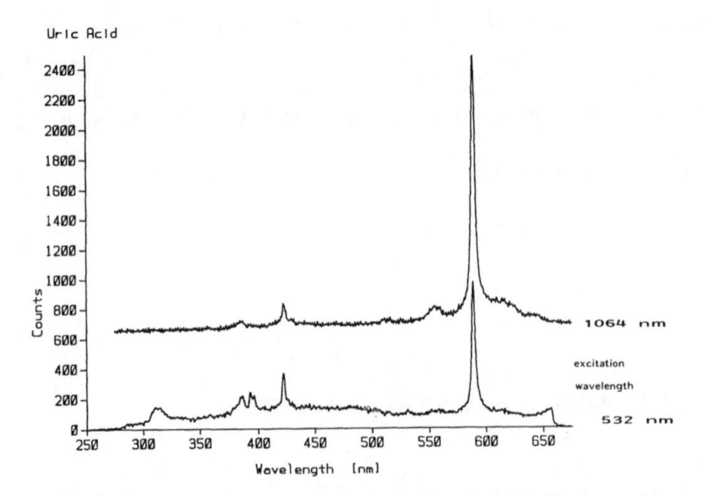

Fig. 5: Comparison of two emission spectra from the same body concrement but with a different excitation wavelengths, the upper spectrum with 1064 nm, the lower one with 532 nm

Thus it can be stated that the typical structures of the emission spectra are independent of the excitation wavelength, whereas, on the other hand the intensities are dependent on it.

4 How Do the Spectra Depend on the Time Delay Between Laser Pulse and Detection ?

Figure 6 shows the time development of the emission spectrum from a uric acid calculus in time increments of 50 ns starting at minus 100 ns before the event. Since the gate time of the OMA is 150 ns . the laser pulse at 532 nm could still be seen in the first four spectra.

In the first 150 ns after the initial laser pulse, the hot plasma with broad emission lines leads to a nearly continuous emission. After this, the plasma has cooled down to such an extent that the emission intensity is still high enough for detection and the lines have become narrow enough to produce a characteristic spectrum. Later the lines only become weaker in intensity. Therefore the best time delay, in order to identify body concrements by their characteristic emission spectrum, is about 100 ns after the laser pulse.

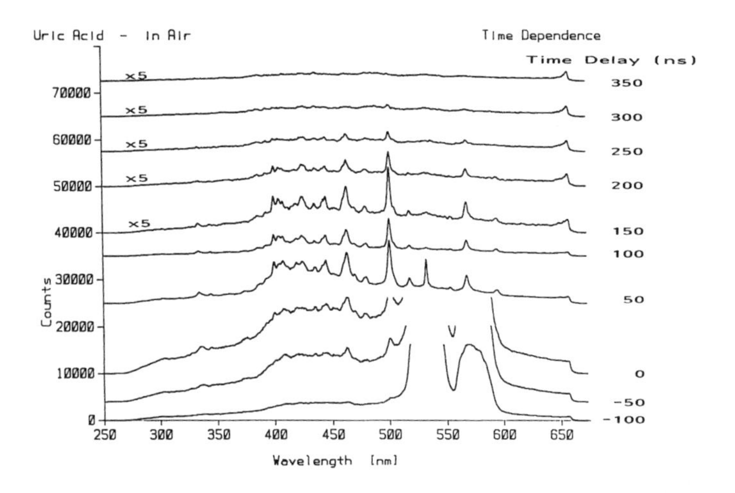

Fig. 6: The time development of the laser induced plasma emission in steps of 50 ns. The initial laser pulse occurred at a time 0. The time delays on the right hand side of each spectrum denote the start of detection

The measurement in figure 6 was done in air because of the much higher intensities achieved in this medium. However, under water the time development is nearly the same. Therefore for our measurements shown in figure 3 - 5 and figure 8 - 13 we used a time delay of 100 ns between the laser pulse and the start of detection.

How Do the Spectra Depend on the Laser Pulse Energy ?

From the same uric acid stone we took five spectra with different energies ranging from 2 mJ to 32 mJ (figure 7). With low energy only a minor information is obtained, but starting with a definite energy - here about 8 mJ - plasma formation takes place. By increasing the energy the plasma will be heated and the emission becomes stronger.

The energy threshold for plasma formation is different for different kinds of stones but is similar to the energy necessary for stone fragmentation and could therefore be a usefull aid in controlling laser lithotripsy.

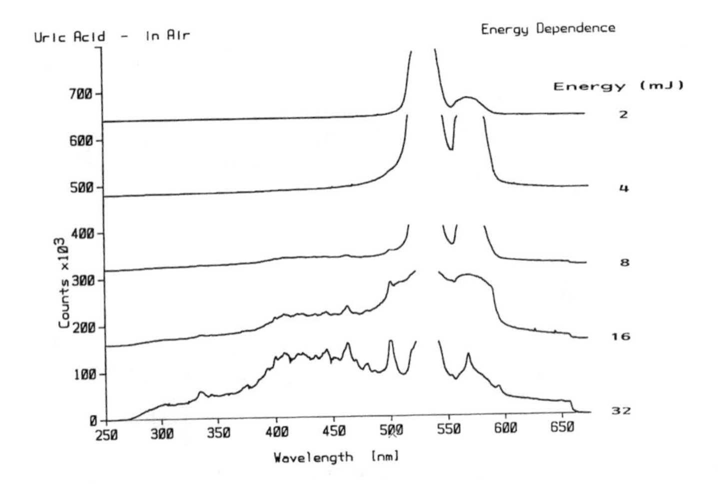

Fig. 7: Energy dependence of the emission spectrum from a uric acid stone. The energy is shown on the right hand side of each spectrum

6 Identification of Body Concrements

Most body concrements are not pure in their composition but conglomerates. As a result we compared standard spectra obtained using pure compounds with those obtained using body concrements. For L-cystine the standard spectrum is shown in figure 8 and it is nearly identical with the respective spectrum in figure 3.

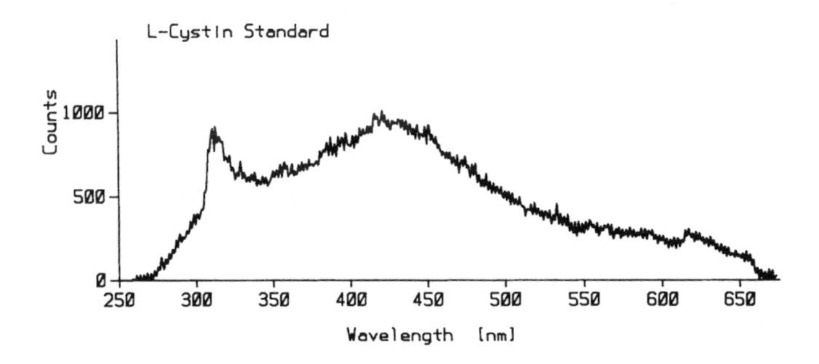

Fig. 8: Standard emission spectrum from a high density pill of pure L-cystine

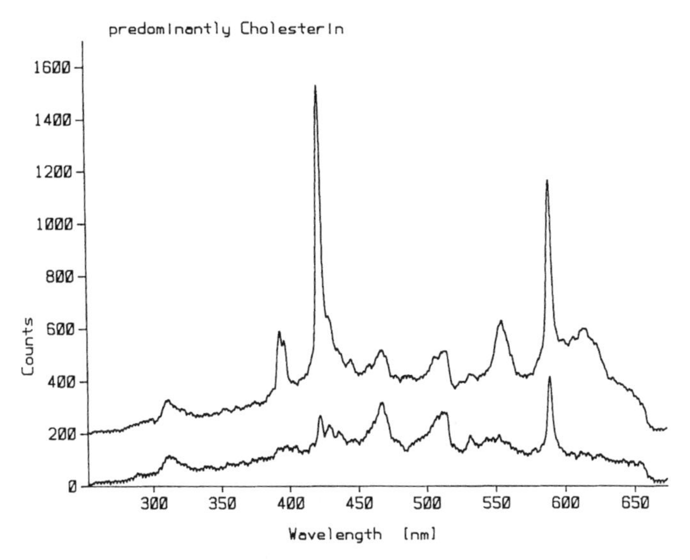

Fig. 9: Spectra of two different biliary stones consisting predominantly of cholesterol

The same comparison is done for cholesterol. Figure 9 and 11 show the spectra of 4 predominantly cholesterol calculi and figure 10 the spectrum of the cholesterol standard.

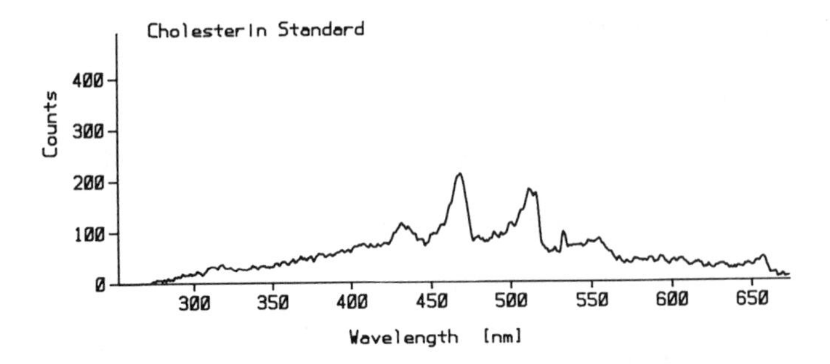

Fig. 10: Standard emission spectrum from a high density pill of pure cholesterol

A conclusion of this is that the characteristic emission of cholesterol could be seen in the spectra of figure 9 as well as some additional emissions. But in figure 11 no cholesterol emission appears because the emission spectroscopy gives, in contrast to the x-ray diffraction analysis not a integral analysis of the whole stone, but a local analysis of this part of the stone which is hit by the laser pulse.

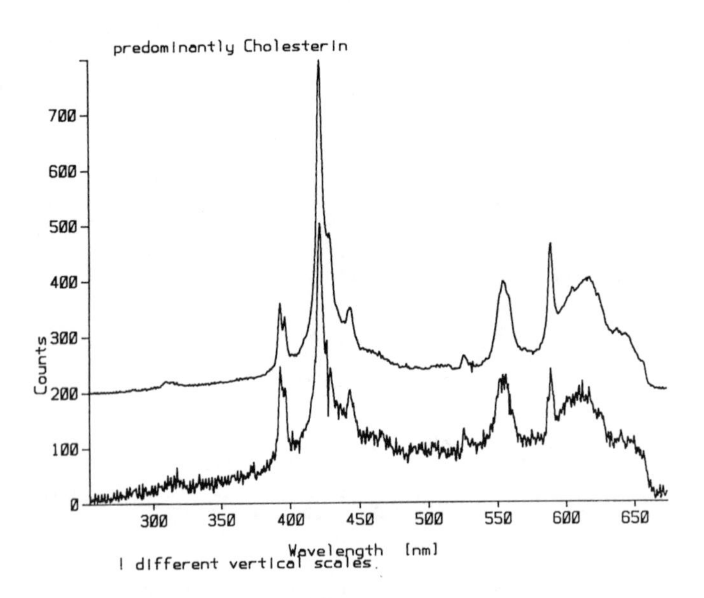

Fig. 11: Spectra from the surface of two different biliary stones consisting predominantly of cholesterol

Fig. 12: Spectra from 3 different calcium-phosphate compounds containing urinary calculi

By comparing the spectra of figure 11 with the spectra of calcium phosphate compounds containing calculi (figure 12) they appear nearly identical because often cholesterol stones have an outer shell of calcium compounds.

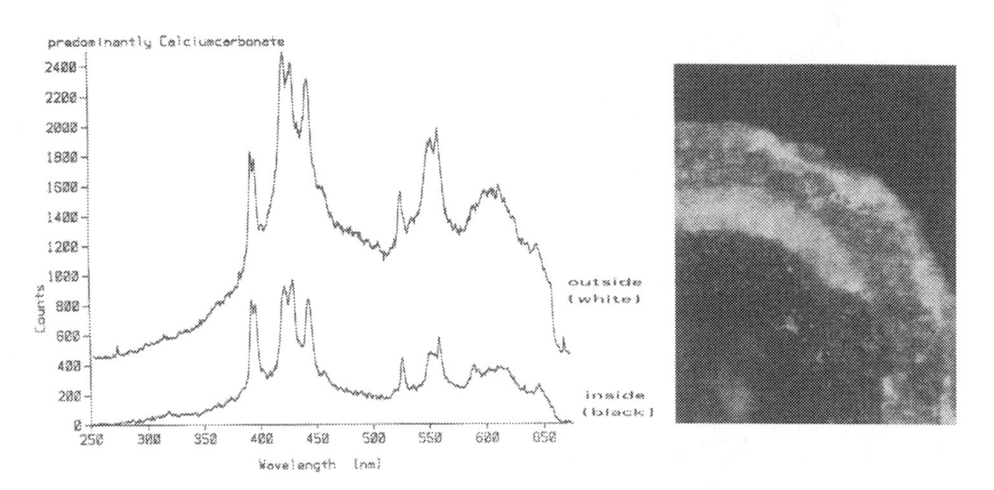

Fig. 13: Spectra from a concrement with two visibly different parts: a black inner part (lower spectrum) and a nearly white outer shell (upper spectrum) as seen in the picture on the right hand side

In figure 13, two spectra of an interesting stone are shown. It is predominantly a calcium-carbonate stone which also consists of impurities of pigments. While the pigments in the outer shell are oxidized by oxygen from the air and are therefore nearly white, the core is still black due to the pigments. The spectra in figure 13 for these different parts of this stone show that the emission spectra are not affected by this difference.

7 Conclusion

We studied the laser induced plasma emission spectra of several human body concrements and standard pills. Considering the experimental parameters which are discussed in this paper it is possible to identify body concrements. Furthermore it is possible to get a feed back from the process during laser lithotripsy. This can be used to control and optimize this process (fig. 14).

The advantage of this method compared to other kinds of stone analysis would be the possibility to use it intracorporally and to obtain an local, not an integral, analysis.

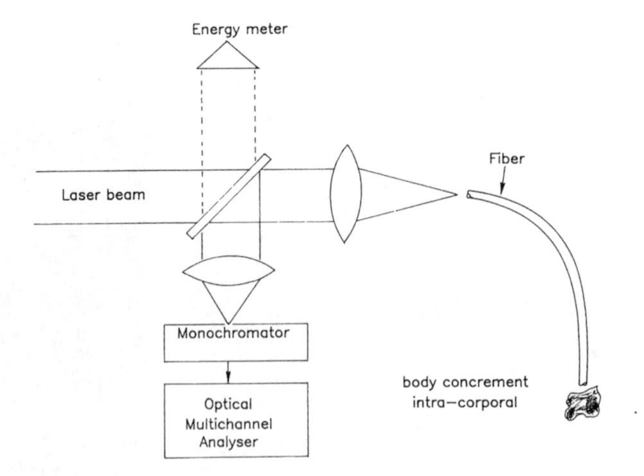

Fig. 14: Setup for laser lithotropsy and a parallel monitoring of the process parameters and stone characteristics by analysis of the emission spectra

This investigation was supported by the BMFT-Grant No. FKZ 13 N 55 11.

1. N.S. Nishioka, P. Teng, T.F. Deutsch, R.R. Anderson: Mechanism of laser induced fragmentation of urinary and biliary calculi, Laser in Life science 1(3), 1987, pp. 231-245

Identification of Biliary and Urinary Calculi by Optical Spectroscopy Compared to X-Ray Diffractometry

W.v. Waldhausen[1], *R. Fitzner*[2], *C. Scholz*[3], *W. Borning*[3], *T. Brodzinski*[4], and *J. Helfmann*[4]

[1]Urological Clinic and Policlinic of the University Clinic Charlottenburg
[2]Institute for Clinical Chemistry and Clinical Biochemistry of the University Clinic Steglitz
[3]Department of Biomedical Technology/Laser Medicine of the University Clinic Steglitz
[4]Laser-Medizin-Zentrum GmbH Berlin, Krahmerstr. 6–10, D-1000 Berlin 45, Germany

The goal of these preliminary investigations is to examine human concrements in relation to intracorporal stone analysis, and in relation to optical properties for laser coupling.

The plots of ten different human concrements, demonstrated in the poster, shown in the appendix, are discussed by the example of a calciumoxalatemonohydrate stone:

Röntgen diffractometry is an established method to identify stones outside the body in clinical chemistry.

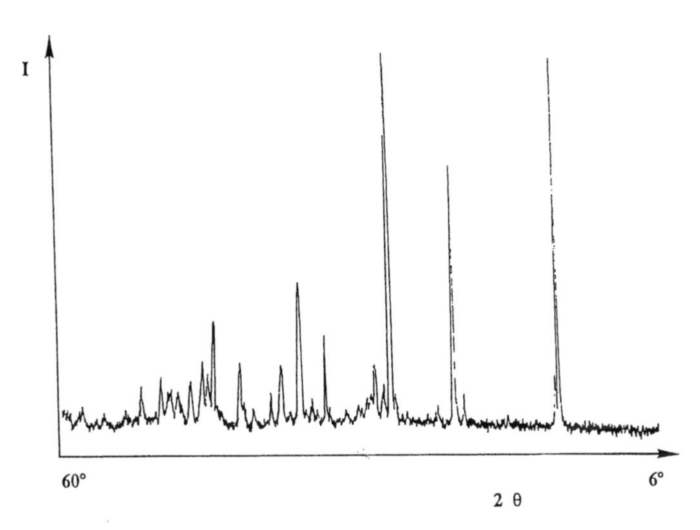

Fig. 1 X-ray diffraction measurement of Calciumoxalatemonohydrate

X-rays striking a crystal interact with the atoms creating waves of scattered X-rays that interfere each other in certain directions. The intensity of the scattered parts of beam is measured with a manoeuvrable detector head (Diagram 1). This method has the disadvantage that it requires a fragment, the size of a cherry stone, which must be pulverised and introduced into the analysis equipment.

The method of the infrared spectroscopy is also applied and suitable for extracorporal analysis of stones. An oscillation frequency, typical for every molecule, can be recognized, especially in the finger print area with a wave number of 1400 - 400 cm^{-1} (Fig. 2).

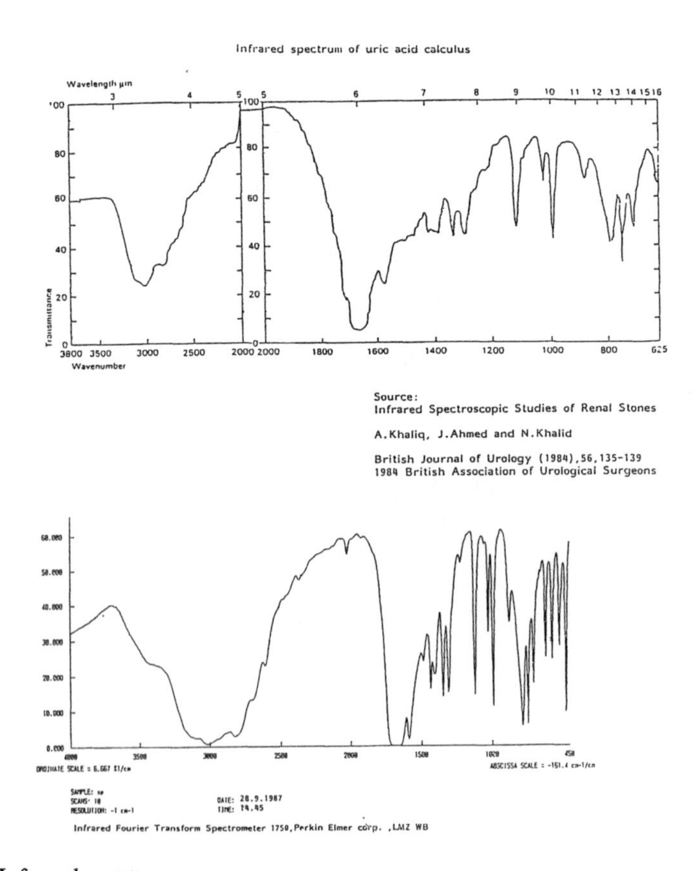

Fig. 2 Infrared spectrum

The identification of body concrements by intracorporal spectroscopy of laser induced plasma with a help of a stimulating laser and an optical multichannel analyser is shown in the paper by J. Helfmann et al. "Identification of body concrements by Fast-Time-Resolved Spectroscopy of Laser Induced Plasma"

Using the Universal Microscope Photometer [UMSP] optical spectra were measured with the absorbed, transmitted, reflected- and scattered fractions (Fig. 3).

- On the one hand from thin sections (20-30 lm thick) by transmission:

  In the absorption spectrum, in transmission, the absorption is plotted against wavelength. It is only partially corrected for reflection (shiny part) and not corrected for scattering (Fig. 4).

- And on the other hand from the stone surface by reflection.

  In the absorption spectrum, in reflection, is the absorption also plotted against the wavelength. It is not corrected for transmission and scattering (Fig. 5).

With only qualitative conclusions one can see the following:

With regard to the transmission spectra, one can see which parts of the radiation remain in the stone (due to absorption, not corrected for reflection and scattering).

The reflection spectra show which part of the spectrum is reflected or back scattered.

Between 400 and 540 nm one can deposit more radiation energy in the stone than at 1060 nm (Fig. 4 and 5). From 600 to 1000 nm the absorption decreases and the reflection increases (Fig. 6 and 7, shown with an example for a Calciumoxalatemonohydrate stone see above). The transmission spectrum of the urinary stones clearly shows between 400 and 540 nm that towards the region of shorter wavelength, absorption and scattering increases in the basic material and more light penetrates the stone.

In comparison to the reflection spectra this effect becomes less towards the shorter wavelength through an increase in the degree of reflection in the same wavelength area. This reflection behaviour can be seen in the spectra of all the ten stones.

In practice, the optimal wavelength with regard to the energy coupling, lies between 500 and 550 nm.

By way of conclusion, these optical spectra, made with an UMSP, and the statement made by this can be a useful method of deciding which laser wavelength must be used to destroy the stone with different physical effects. However it cannot be used to identify it.

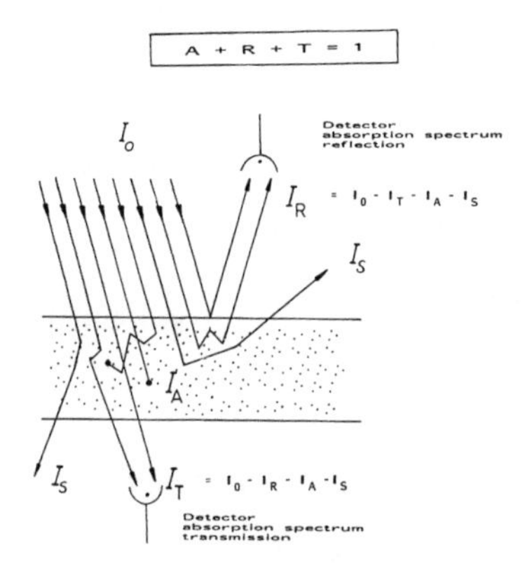

Fig. 3 Experimental conditions for the UMSP

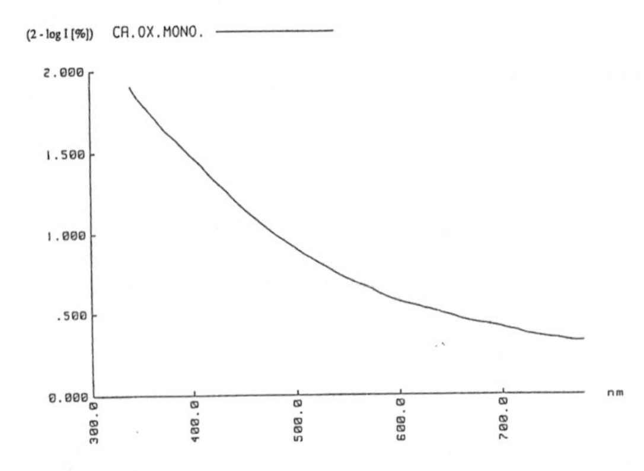

Fig. 4 UV/VIS absorption spectrum transmission

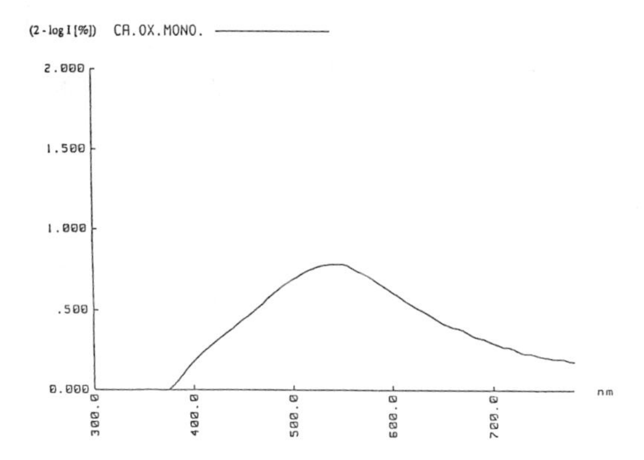

Fig. 5 UV/VIS absorption spectrum reflection

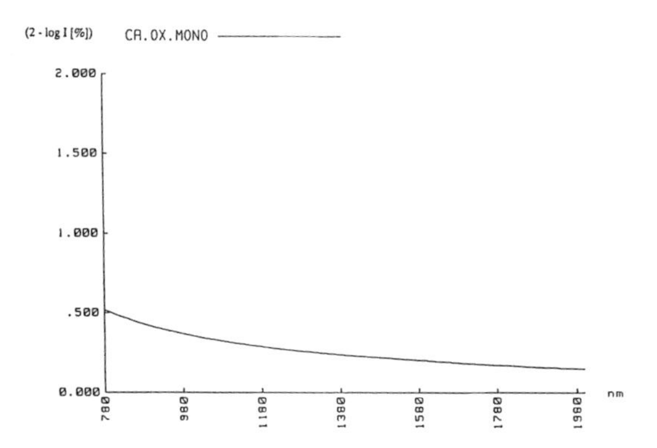

Fig. 6 NIR absorption spectrum transmission

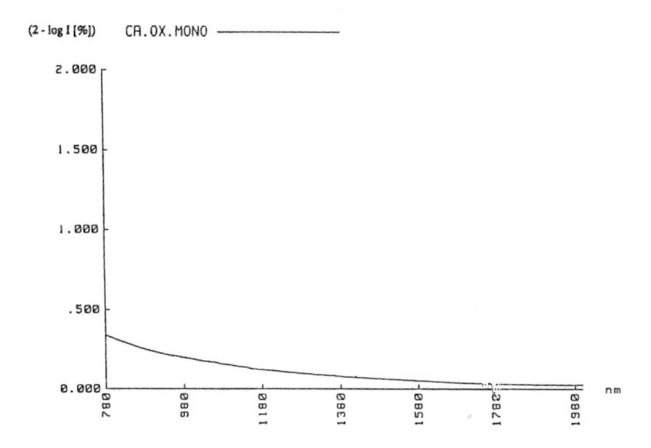

Fig. 7 NIR absorption spectrum reflection

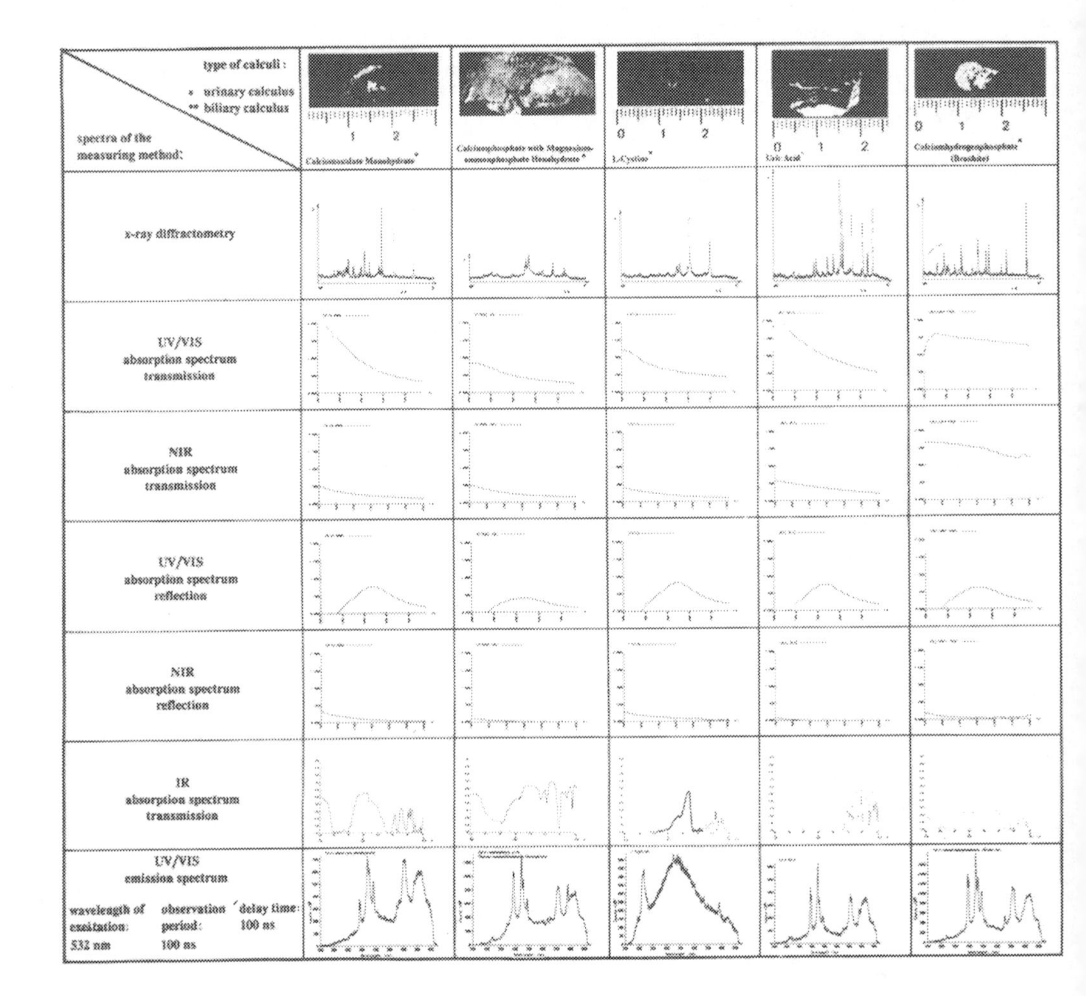
type of calculi :
* urinary calculus
** biliary calculus
spectra of the measuring method:
x-ray diffractometry
UV/VIS absorption spectrum transmission
NIR absorption spectrum transmission
UV/VIS absorption spectrum reflection
NIR absorption spectrum reflection
IR absorption spectrum transmission
UV/VIS emission spectrum
waveleugth of exsitation: 532 nm
observation period: 100 ns
delay time: 100 ns

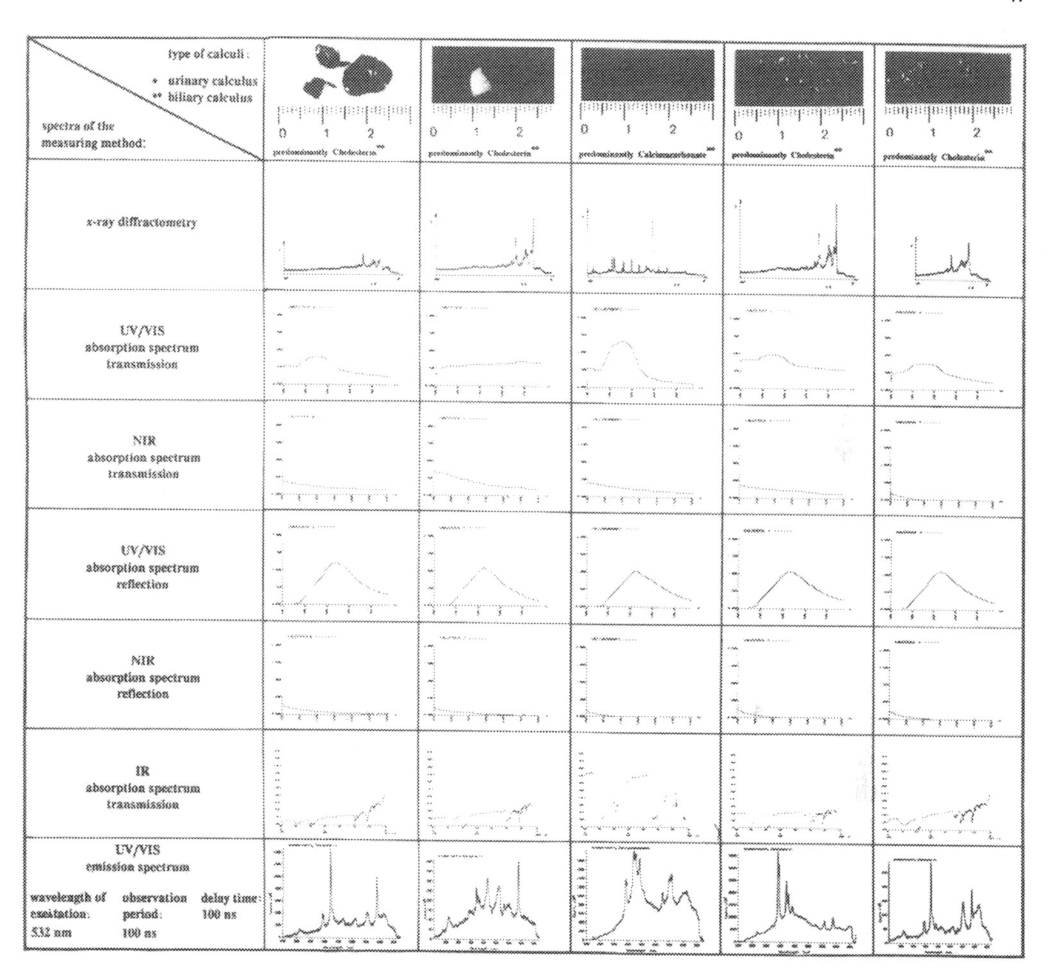

Method:

x-ray diffractometer: PW 1840, PHILIPS
Microscope-Spectral-Photometer UMSP 80,Carl Zeiss

Infrared Fourier Transform Spectrometer 1750,Perkin Elmer corp.

Optical Multichannel Analyser OMA III,EG & G
Excitation laser: Nd-YAG (532 nm) DCR 3,Spectra Physics LPD

Primary Results in the Laser Lithotripsy Using a Frequency Doubled Q-Switched Nd:YAG Laser

K. Dörschel[1], *H.-P. Berlien*[1], *T. Brodzinski*[1], *J. Helfmann*[1], *G. Müller*[1], *and C. Scholz*[2]

[1]Laser-Medizin-Zentrum GmbH Berlin, Krahmerstr. 6–10, D-1000 Berlin 45, Germany
[2]Freie Universität Berlin, Fachgebiet für Biomedizinische Technik, Schwerpunkt Lasermedizin, D-1000 Berlin, Germany

Experiments in using a frequency doubled Q-switched Nd:YAG laser (532 nm) for laser lithotripsy were carried out. First results on the efficiency of fragmentation in using this technique will be presented. In detail the feasibility for the fragmentation of kidney, bladder and biliary stones will be discussed.

The two different laser systems most frequently used until now for stone destruction have been the Nd:YAG laser in Q-switched operation (1064 nm) and the flash lamp pumped dye laser (500 - 600 nm). An interpretation of the results obtained is however problematic as not only the wavelength but also the laser pulse times (Nd:YAG typ. 10 nsec; dye laser typ. 1 μsec) differ from one another.

When investigating the influence of the wavelength during stone destruction, the use of a frequency doubled Nd:YAG laser appears to be ideally suitable, as when the frequency is doubled the pulse length remains constant. As our investigations show, stones can be destroyed with both wavelengths 532 and 1064 nm, whereby at 532 nm the required pulse energy is clearly much less, a typical example of which is shown in Fig. 1.

In the experiments the laser beam is focussed using a lens directly onto the stone surface, which is located under water (Fig. 2). The irradiated area, in focus, was 50 μm x 120 μm. With the laser system used (DCR-3-Spectra Physics, 8 nsec pulse length) the laser wavelength can be switched without any significant changes in the beam geometry.

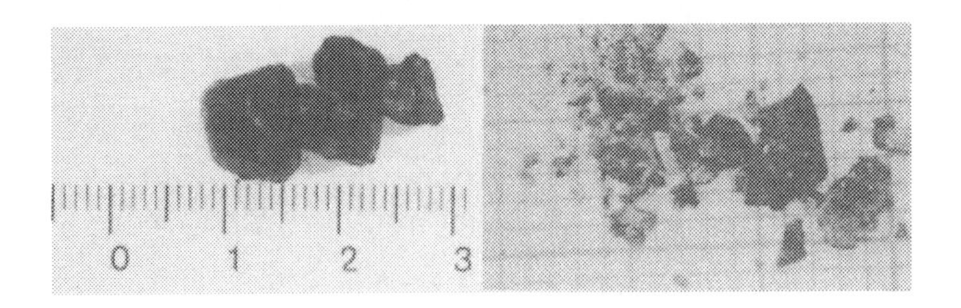

Fig. 1 left: Cholesterine gall stone before laser irradiation
right: Stone concrements after 1500 laser shots 40 mJ, 532 nm, 30 Hz

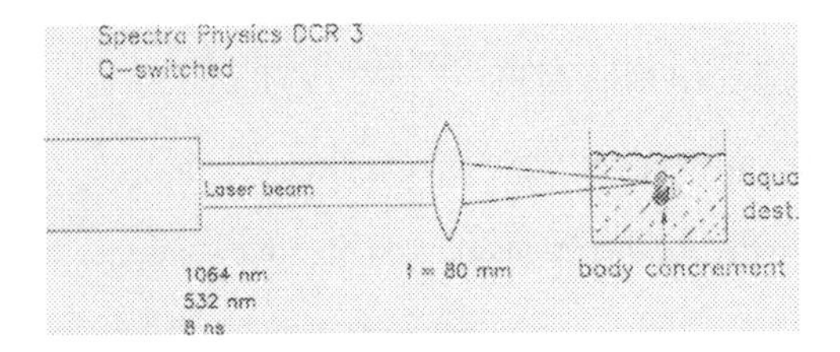

Fig. 2: Experimental set up

A typical result observed by the laser irradiation of stones is the production of dust at a definite pulse energy (532 nm/150 mJ, 1064 nm/800 mJ) independent of the type of stone and start of destruction in the form of small particles (Fig. 4) at a pulse energy which is additionally dependent on the type of stone.

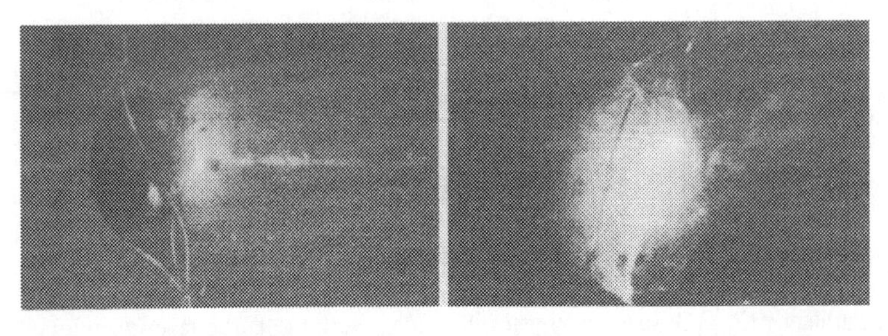

Fig. 3 left: Above definite pulse energies (532 nm/150 mJ) dust formation starts
right: Fragmentation starts at higher pulse energy, which is additionally dependent on the type of stone

The results of these experiments are summarized in Fig. 4.

Type of Calculi		Threshold of Dust		Threshold of Fragmentation	
		532 nm mJ/mm^2	1064 nm mJ/mm^2	532 nm mJ/mm^2	1064nm mJ/mm^2
Urinary Calculi		150	800		
Urid acid:	No. 1			1 700	8 000
	No. 2			300	1 800
	No. 3			1000	
Calciumphosphat Magnesiumammon-phosphat-Hexahydrate	No. 4			600	
Biliary Calculi		150	800		
Cholesterin:	No. 1			1 300	
	No. 2			1 300	
	No. 3			700	4 000
	No. 4			300	
	No. 5			500	
	No. 6			250	1 400
	No. 7			1 300	
Calcium-carbonate	No. 8			300	1 600

Fig. 4: Energy threshold for production of dust and for small particles for different classes of stones

The results show that the destruction threshold for all stone types at 532 nm clearly is much lower than by 1064 nm. If one compares the absorption behaviour of the stone material at 532 nm and 1064 nm, the destruction threshold appears to behave contrary to the absorption coefficient. The large degree of variation of the destruction threshold with the same stone type clearly shows that a series of other factors such as the stone morphology or different constituents have an effect on the fragmentation threshold.

This investigation was supported by the BMFT-Grant No. FKZ 13 N 55 11.

Stone Disintegration with Lasers – A Comparison Between Pulsed Nd:YAG- and Dye Laser

Th. Meier, U. Fink, and R. Steiner

Institut für Lasertechnologien in der Medizin an der Universität Ulm,
Postfach 4066, D-7900 Ulm, Fed. Rep. of Germany

In order to compare the action of different lasers on stones it is necessary to have a measure for the fragmentation power as well as to have some kind of standard stone. We measured the shockwave pressure after a laserinduced breakdown inside and outside of natural and artificial stones. It turned out that this procedure could help to get a deeper insight into the fragmentation process, but it is not suited for a quantatitative study of fragmentation. Instead, we measured the ablation of material directly with an electronical balance.

1. Experimental Procedure

In this comparative study we used a Q- switched Nd:YAG laser (JK Lasers, 1060 nm, 200 mJ, 12 ns) and a flashlamp pumped dye laser (Candela Corporation, 200-600 mJ, 500 ns).

The experimental set-up is shown in Fig. 1.

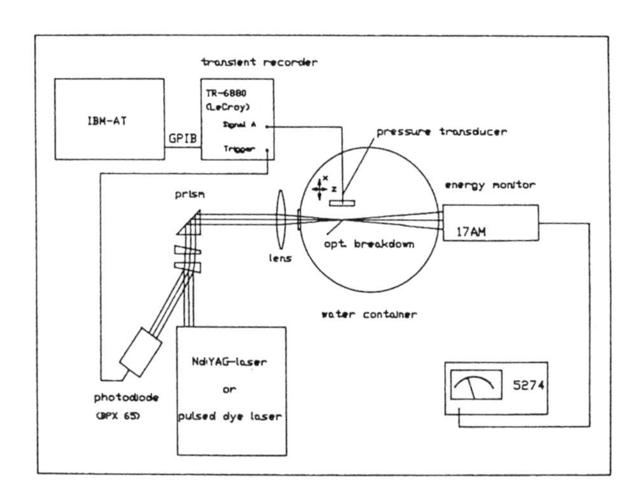

Fig. 1. Experimental set-up. The pulsed laser was focused either in water or on a stone with embedded pressure transducer.

The laser beam was focused by a 50 mm lens onto the stone under in vestigation mounted in a glass vessel filled with water. The electrical signal from the pressure detector was fed to a fast oscilloscope or, alternatively, to a transisient recorder. The detector simply consisted of a piezoelectric foil in a special geometry, which could easily be arranged around the stone or embedded into the stone (Fig.2). Details of the experimental procedure and examples of shockwave recordings in water and stone are described elsewere /1/. Some considerations on detector geometry are discussed in another contribution by the same authors in these proceedings. Stones under investigation were kidney stones, natural limestones, and artificial stones made from plaster or cement.

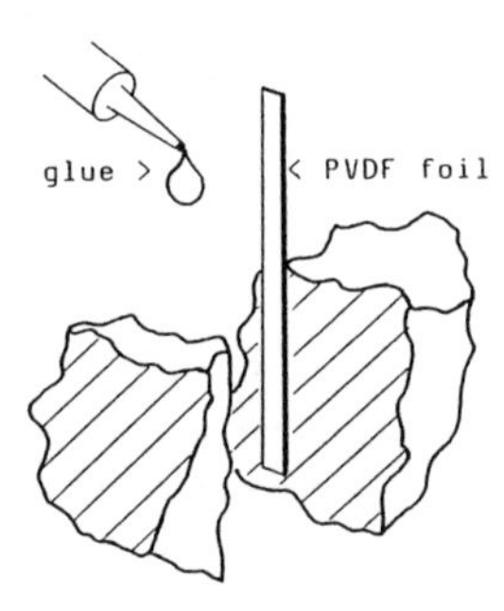

Fig. 2. Preparation method for natural stones like kidney or limestone.

2. Results

Just by visual observation during and after a number of laser shots we could realize principal differences in the action on the stone by different lasers. With the Nd:YAG laser we always observed an optical breakdown in water (with or without stone) but not with the pulsed dye laser. This is due to the shorter pulse length and better beam quality of the ND:YAG compared to the dye laser. To induce something like an optical breakdown with dye laser it is necessary to seed the water with some solid material like e.g. polishing powder. Correspondingly, on limestone we found after a number of shots with the Nd:YAG laser narrow holes (appr. 0.2 mm in diameter) drilled into the stone while the dye laser produced broader and more irregular cavities (On the more brittle parts of the kidney stones this difference is less pronounced).

How are these findings correlated to the shockwave measurements? Figure 3 shows a recording of a shockwave propagating inside a limestone (upper curves). We were not able to detect any shockwave signal.

However, at these conditions fragmentation of the softer stones was even better.

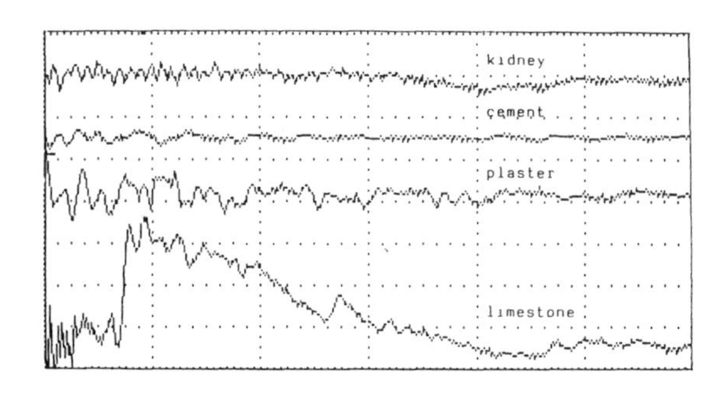

Fig. 3. Shockwave pressure recordings in different stones. The dye laser was focused directly on the surface, the detector was embedded app. 3 mm from the surface. x-axis: 500 ns/div.; y-axis: 100 mV/div., corresponding to app. 30 MPa/div.

As the recordings with the dye laser show this is not a question of wavelength (Fig.4).

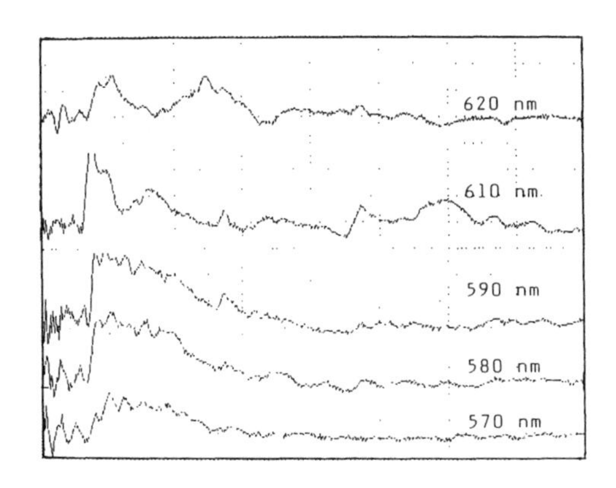

Fig. 4. Shockwaves in limestone generated by the dye laser at different wavelengths. (other experimental parameters as in Fig.3).

In order to study the wavelength dependence of the fragmentation more directly we recorded the weight of the stones during 10 laser shots of app. 300 mJ. In the case of the dye laser we plotted the mean mass-loss against the wavelength (Fig. 5) for different stone types.

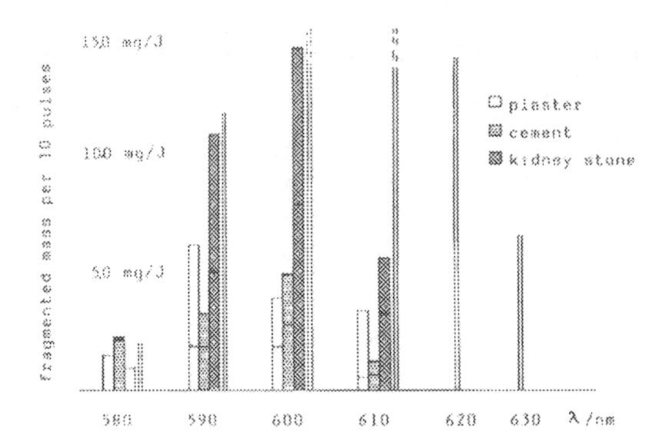

Fig. 5. Mass-loss per energy against the wavelength of the dye laser for various stones. The blank narrow bars represent data obtained with a 600 um PCS fibre.

Although there are strong variations in two independent sets of measurements (indicated by the horizontal lines in the bars of Fig. 5) we can observe a maximum in the red wavelength regime of dye (Rh 6 G). For comparison we also included the results obtained by transmitting the laser pulses via a 600 um fibre. It seems that this method is more effective than focusing the beam with a lens on the stone. Extensions of these measurements to shorter and longer wavelengths show that for Rhodamine 700 (720-780 nm) the fragmentation is comparable to the results with Rh 6G but reduced at 520-580 nm with Coumarine 102.

The results for the Nd:YAG laser are not shown in Fig. 5. The fragmentation rate of the Nd:YAG laser under the same experimental conditions is a factor of ten lower than that obtained with the dye laser (some mg/J).

3. Conclusion

Shockwave measurements inside the stone can help to understand the fragmentation process. It turns out that the existence of shockwaves deep in the stone means little fragmentation, i. e. hard stones, while good fragmentation can be observed in softer stones without detectable shockwaves. The fragmentation process in the case of the Q-switched Nd:YAG laser is mainly based on the generation of a laser-induced breakdown in water and by coupling the shockwave onto the stone. With the dye laser the exploding plasma is generated directly on the stone surface which may lead to a more efficient coupling of mechanical energy to the stone. The wavelength dependence of the fragmentation is of less importance than expected. Possibly, the role of the wavelength is restricted to the ignition phase of the plasma and, by that, it may be responsible only for the fragmentation threshold.

References:

1. Th. Meier, E. Keckeis, R. Steiner: In Proceedings of the 7 th Congress of the International Society for Laser Surgery and Medicine, Munich (1987), ed. by W. Waidelich, in print (Springer, Berlin, Heidelberg)

Gallstone Lithotripsy Nd:YAG Laser Versus Dye Laser

H. Wenk[1], *J. Pensel*[2], *P. Hering*[3], *Th. Dann*[1], *and G. Baretton*[4]

[1]Department of Surgery, University of Lübeck, D-2400 Lübeck, Fed. Rep. of Germany
[2]Medical Laser Center Lübeck, D-2400 Lübeck, Fed. Rep. of Germany
[3]Max-Planck-Institut für Quantenoptik, D-8046 Garching, Fed. Rep. of Germany
[4]Institut of Pathology, University of Lübeck, D-2400 Lübeck, Fed. Rep. of Germany

Introduction

The development of a pulsed Nd:YAG Laser in Q-switched mode allowed the mechanical desintegration of concrements. After successful in vitro studies there were done animal experiments. The results, which are reported recently, showed no severe soft tissued reactions and were the justification for clinical use (2).

The available transmission systems allow the use of the laser system in modern flexible endoscopes. However, application in the bile duct is difficult, for the optomechanical couple has a diameter of 2.2 mm and the 600 um quartz guide is rigid. In spite of performing a large papillotomia, it is much more difficult to place the system into the bile duct than into the pancreatic duct.

These difficulties were the reason for other experiments about shock wave lithotripsy with another laser system.

Material and methods

1. in-vitro-experiments

By a flash lamp plused dye laser with a wavelength of 590 nm human gall stones were destroyed. The pulse energy of 40 mJ allows a spontanious breakdown and a renunciation of an optomechanical couple. The transmission fiber has a diameter of only 200 um and is much more flexible than the fiber of the Nd:YAG laser.

	Nd:YAG LASER	DYE LASER	
wavelength	1064	590	nm
pulseduration	12	1500	ns
pulse-energy	40	40	mJ
transmission	600	200	um

2. Animal experiments

Animal experiments were done with female pigs to examine the tissue reactions of the bile duct and the gallbladder.
The tissue was explanted immediately, after one week, two weeks and three months.

Results

1. in vitro experiments
All stones were destroyed in short time. We used in single shot mode between 10 and 1188 pulses.

GALLSTONE LITHOTRIPSY (DYE LASER)

weight (mg)	400	7200	790	4500	610	3300	1000	2190
main const.	CHOL	CHOL	CHOL	CHOL	CHOL	CHOL	CHOL	CHOL
pulses	13	1188	72	420	43	39	10	20

2. Animal experiments
The immediately explanted tissue showed fresh bleeding and beginning necrosis of the whole wall.
1 week after lithotripsy we saw a sharply demarked necrosis and demarkation by granulation in the subserosa.
2 weeks after laser therapy the muscle layers showed a reparative fibrosis.
3 months after lithotripsy no residuum of the treatment could be observed. There was no stenosis by scarification.
In one case we observed a perforation of the bile duct into the portal vein.

Discussion and conclusion

In contrast to the tissue reactions by using the Nd:YAG laser a total necrosis of the bile duct's wall by the dye laser in equipotent application is recognized. The destruction of the tissue and bleeding suggest mechanical effect. Thermic effect could not be recognized when scrutinizing the laser lithotripsy with a thermocamera.

We did not loose any animals intra- or postoperatively in spite of transmural necrosis. In the explanted tissue we saw neither complete perforation nor peritonitis as a sign of penetration.

In spite of the positive result of the long time experiment and other descriptions in the literature (1) our results order to be careful with clinical use of the dye laser system. It must be examined, which modifications of the system allow a quick lithotripsy with spontane breakdown and prevent a transmural necrosis of the tissue.

The favourable circumstances - thin transmission systems, renunciation of a couple, no requirement of a rinsing tube - have avoken great expectations into the dye laser. Nevertheless, the Nd:YAG laser is the instrument of first choice in present time.

Literature

1. **G. Watson, St. Murray, St. Dretler, J. Parrish:**
   Am Journal of Urology 138, 199-202 (1987)

2. **H. Wenk, V. Lange, F.W. Schildberg, A. Hofstetter:**
   Laser in medicine and surgery 3, 194-200 (1987)

Acute Shock-Wave Induced Effects on the Microcirculation +

A.E. Goetz[1], *R. Königsberger*[1], *F. Hammersen*[2], *P. Conzen*[1], *M. Delius*[1], *and W. Brendel*[1]

[1]Institut für Chirurgische Forschung, Ludwig-Maximilians-Universität, Klinikum Großhadern, Marchioninistr. 15, D-8000 München 79, Fed. Rep. of Germany

[2]Anatomisches Institut, Technische Universität, Biedersteinerstr. 29, D-8000 München 40, Fed. Rep. of Germany

Introduction

The extracorporeal shock-wave lithotripsy (ESWL) of kidney - and gallbladder stones has been initiated at the Institute of Surgical Research /1,2,3,5/. Meanwhile it is a clinically established routine method and more than 1 million patients suffering from kidney calculi and 200 patients with gallbladder stones have been treated successfully /20/.

With ESWL of kidney calculi the following side-effects have been reported in a small percentage of the patients: bruising of skin, a transient macro- and microhematuria, hematomas in the kidney capsule, ureteric colics, a reduction of the effective renal plasma flow, perirenal fluid accumulation and a transient increase of serum enzyme levels of LDH, GOT, CPK and NAG /5,6,10,18/. These findings induced us to evaluate directly the acute effects of shock waves on the microcirculation in an in-vivo experimental approach. In contrast to the procedure during stone desintegration the tissue was centered into the focus of the shock-wave.

Methods

Transparent double-frame aluminium chambers were adapted to the dorsal skin fold of male Syrian golden hamsters weighing about 80 g /12,15,16/. Using this preparation microscopic access to the skin-muscle microcirculation was achieved. 2 - 3 days after recovery from microsurgery and anesthesia animals were anesthetized by pentobarbital and fixed in a transparent water-tight plastic tube which was suffused by a mixture of oxygen and air (fraction of oxygen: 0,4). The chamber preparation projecting over the tube via a lenghtwise running slot in-

---

+ Supported by Kurt Körber Foundation

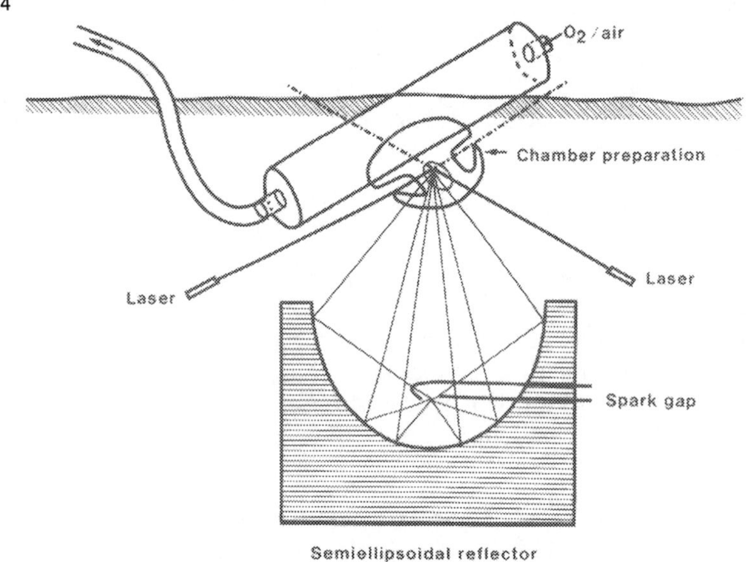

Fig.1. Scheme of shock wave generation with the Dornier lithotripter and of positioning the transparent chamber preparation in the second focus of the shock wave by the intersection of 2 laser beams.

to the warmed water bath (35-36 C) was centered horizontally in the second focus of the rotationally symmetric semi-ellipsoidal reflector by the use of 2 intersecting laser beams (Fig.1). The intact epidermis was directed towards the discharging electrode.
The shock wave generator used in these experiments and its physical properties were described earlier (9,14). Using the Dornier Lithotripter XM1 high energy shock-waves were generated in the geometrical focus of the semi-ellipsoidal reflector by a high current underwater spark discharge. The shock wave is propagated spherically and reflected by the wall of the ellipsoid and concentrated in the second focus. The characteristics of these shock waves were: A steep pressure slope in the nanosecond range, a pulse duration of about 1 microsecond, a peak pressure up to 1000 bar in the "second" focus and a peak pressure field of about 1.5 cm in diameter /9/. For evaluating microvascular and vascular diameters by fluorescence photomacroscopy and fluorescence-videomicroscopy FITC-dextran (150 000 dalton) was administered intravenously. Randomly selected areas within the skin muscle microcir-

culation were photographed prior to and intermittently up to 10 minutes after shock-wave application using the Wild MPS 55/51 photoautomat (Leitz, Wetzlar, FRG). For microhemodynamic measurements microvascular areas were continuously visualized by fluorescence videomicroscopy by a modified Leitz Orthoplan-microscope (Leitz, Wetzlar, FRG). The microscopic images were transferred by a high sensitivity TV camera (C 2400, Hamamatsu Co., Japan) to a TV monitor and stored on videotape (SONY Co., Japan). Later these images were analyzed off-line at magnifications of 1200x from the TV screen. 42 hamsters were randomly assigned to the control group (n=7) or to the groups receiving 1 (n=12), 10 (n=12) or 100 (n=12) discharges at a voltage of about 15 kv and a capacity of 90 nF. Single preparations were subjected to superfusion fixation for electronmicroscopy after exposure to one shock wave.

Results

In the control group no significant microcirculatory changes could be measured. Following one discharge the microcirculatory changes observed in the peak pressure area were as follows: A transient vasoconstriction of the consecutive arteriolar microvascular segments, i.e. of transverse arterioles, terminal arterioles and precapillaries (Fig.2 and 3). Luminal diameters of postcapillaries and venules did not change significantly. Arteriolar vasoconstriction was accompanied by transient stasis in the constricted vessels and in downstream microvessels, i.e. in capillaries, postcapillaries and venules. Continuous videorecordings showed a maximum vasoconstriction of the arteriolar segments at 30 seconds lasting for 3 to 7 minutes.
The findings described below were increased with the numbers of shock applied to the intact tissue.
Microhemorrhages were detected at the site of small venules. Fluorescence macroscopic images revealed leakage of the macromolecules (FITC-dextran 150,000) with special regard to small venules and postcapillary segments. Moreover amorphous conglomerates were observed in the intravascular space of venules. Intravenously administered acridine orange, a substance which binds to platelets in vivo /21/, revealed that the conglomerates mainly consist of platelet aggregates.
Thus venular blood flow was impeded and occasionally venular stasis occured even if reperfusion following the arteriolar vasoconstriction was already present.
Electronmicrographs of venules and postcapillaries revealed endothelial ruptures and leukodiapedesis after one shock wave only. In contrast arterioles were still intact.

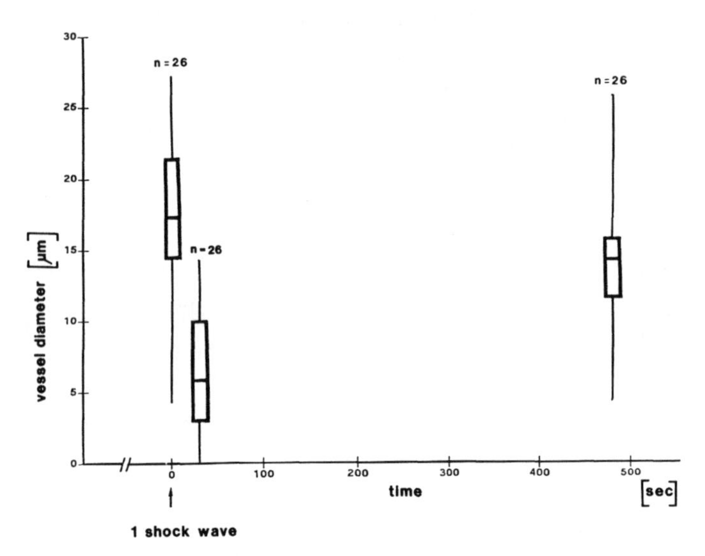

Fig.2. Changes of vessel diameters in transverse arterioles after 1 shock wave are given in box plots.

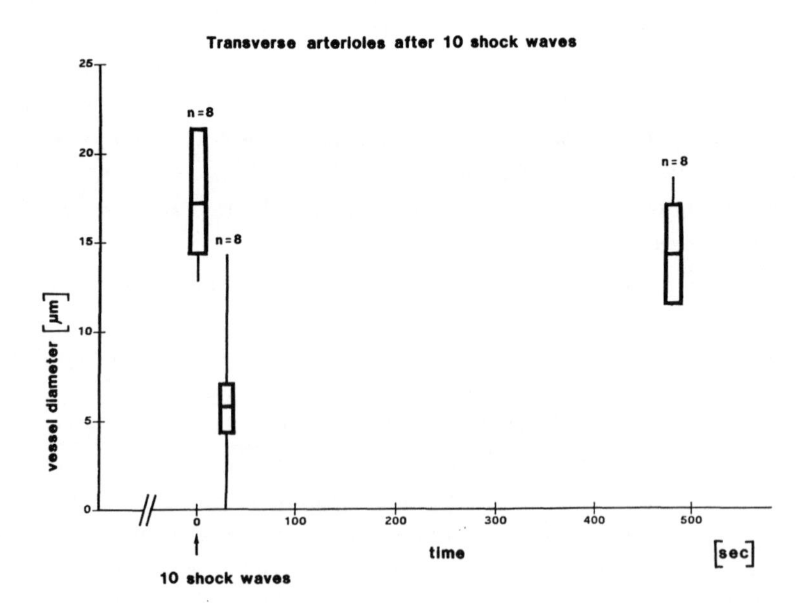

Fig.3. Changes of vessel diameters in transverse arterioles after 10 shock waves are given in box plots.

Discussion

Microcirculatory changes after shock-wave treatment studied in the skin muscle microcirculation of anaesthetized Syrian golden hamsters explained in part the clinical findings and symptoms mentioned initially /5,6,18/. In addition, the results correlated to recent histologic findings in canine kidneys and lungs following ESWL /9,10/. So far these effects may be attributed to a limited contusion of the immature parenchyma in the peak pressure area. The exact mechanisms responsible for the microcirculatory alterations are still discussed controversially.

Concerning the arteriolar vasoconstriction after shock wave application, a metabolic mechanism or a release of mediators, probably of the eicosanoid cascade, or a pressure/tension sensitive response of the smooth muscle cells within the arterioles has to be considered /4/.

As to the effects at the venular site evidence exists that the phenomenon of acoustic cavitation /7/ might be one of the proximate causes inducing the tissue damage with ESWL.

Living systems have been reported to possess cavitation nuclei /17/. We observed that underwater gas bubbles were generated in the "second" focus in the degased water bath during discharge and rose to the water surface. Multiple spots occurred in the teflone membrane covering the preparation area after one shock-wave, when the teflone membrane was directed towards the discharging electrode. These spots were indicative of jet production induced by collapsing cavitation bubbles /19/. Moreover evidence of liquid jet impacts and the observation of cavitation bubble collapse using X-ray films, thin aluminium sheets and thick metal plates have been described /7/. Recent studies using ultrasound imaging during shock wave exposure to the liver demonstrated a transient reflex in liver veins. Moreover these transient reflexes were transported away with the blood stream in this vessel (Delius M. et al., unpublished results). These results suggest that caviation indeed might exist in-vivo. In addition cavitation induced biological effects with ultrasound are comparable to those induced by shock wave exposure, e.g. platelet aggregation and stasis of blood flow /8,11,13/.

Considering the present results for ESWL of kidney- and gall-bladder calculi it should be taken into account that the clotting system of the patient has to be well controled prior to shock-wave treatment. Our findings suggest that ESWL should not be performed if there is any suspicion for extremely liable tissue (inflammation, neovascularization) in the focus area during treatment.

References

1. W.Brendel, C.Chaussy, B.Forssmann, E.Schmiedt: Br. J. Surg. 66 , 907 (1979

2. W.Brendel, G.Enders: Lancet i , 1054 (1983)

3. W.Brendel, M.Delius, G.Enders: In Enterohepatic circulation of bile acids and sterol metabolism, ed. by G. Paumgartner, A. Stiehl, W. Gerok (MTP Press, Lancaster, Boston, The Hague, Dordrecht 1983) p. 381

4. M.E.Burrows, P.C.Johnson: Am.J. Physiol. 245 , H796 (1983)

5. C.Chaussy, W.Brendel, E.Schmiedt: Lancet ii , 1265 (1980)

6. C.Chaussy, E.Schmiedt, D.Jocham, W.Brendel: J. Urol. 127 , 217 (1982)

7. A.J.Coleman, J.E.Saunders, L.A.Crum, M.Dyson: Ultrasound Med. Biol. X 13 , 69 (1987)

8. L.A.Crum: Proc. Ultrasonics Symp. 1 , 1 (1982)

9. M.Delius, G.Enders, Z. Xuan, M.Rath, G.Liebig, W.Brendel: Utrasound Med. Biol. 13 , 61 (1987)

10. M.Delius, G.Enders, Z.Xuan, H.G.Liebig, W.Brendel: Ultrasound Med. Biol. (in press)

11. M.Dyson, J.B.Pond, B.Woodward, J.Broadbent: Ultrasound Med. Biol. X 1 , 133 (1974)

12. B.Endrich, K. Asaishi, A.Goetz, K.Meßmer: Res. exp. Med. 177 , 125 (1980)

13. L.A.Frizzell, D.L.Miller, W.L.Nyborg: Ultrasound Med. Biol. 12 , 217 (1986)

14. B.Forssmann, W.Hepp, Ch. Chaussy, F.Eisenberger, K.Wanner: Biomed. Techn. 22 , 164 (1977)

15. A.Goetz, B.Endrich, C.Laprell, K.Meßmer: Bibl. Anat. 20 , 65 (1981)

16. A. Goetz, M.D. Thesis, University of Munich (1987)

17. E.N.Harvey, D.K.Barnes, W.D.McElroy, A.H.Whitely, D.C. Pease, K.W.Cooper: Cell Compl. Physiol. 24 , 1 (1944)

18. J.V.Kaude, C.M. Williams, M.R.Millner, K.N.Scott, B.Finlayson: Am. J. Roentg. 145 , 305 (1985)

19. C.F.Naude, A.T.Ellis: Trans.Am.Soc.Mech.Eng., Ser. D., J. Basic Eng. 83 , 648 (1961)

20. T.Sauerbruch, M.Delius, G.Paumgartner, J.Holl, O.Wess, W.Weber, W.Hepp, W.Brendel: New Engl. J. Med. 314 , 818 (1986)

21. G.J. Tangelder, D.W. Slaaf, R.S.Reneman: Thrombosis Res. 28 , 803 (1982)

Effects of Laser Pulses on Cells and Tissue

G. Dohr[1], *H. Schmidt-Kloiber*[2], *E. Reichel*[2], *and H. Schöffmann*[2]

[1]Department of Histology and Embryology, University of Graz, A-8010 Graz, Austria

[2]Department of Experimental Physics, University of Graz, A-8010 Graz, Austria

1. INTRODUCTION

Laser-induced shock wave lithotripsy with a Q-switched Nd-YAG laser was developed in Graz /1,2/ and, after thorough preliminary experiments, has been successfully introduced into clinical care. In vivo studies on pigs /3,4/ and in vitro studies on rat organs /5,6/ and human cell cultures have been done to investigate the effects of laser pulses on biologic material.

Laser-induced shock waves arise in the immediate vicinity of a ureteral calculus or gallstone. The effects of these shock waves on the surrounding cells, tissues and organs can be studied.

Tissue injury is unusual in clinical practise because laser lithotripsy is done under visual control with optical fiber equipment. Nevertheless, we wanted to evaluate the tissue damage if the laser-induced breakdown or shock waves directly struck the surface of an organ. This paper presents light and electron microscopic findings of in vitro experiments on isolated rat organs and human fibroblast cell cultures.

2. MATERIAL and METHODS

The rats were drugged and their kidneys removed. Each organ was mounted on a prepared plate, submerged in saline warmed to 37°C and brought into the focus of the laser beam. After being exposed to one or more laser pulses, the organ was removed from the chamber and fixed with 3% glutaraldehyde (GA) and 2% p-formaldehyde (PFA) in phospate buffered saline (PBS) at pH of 7,4. Specimens for electron microscopy were additionally fixed with 1% OsO_4. After rinsing and dehydration, the specimen was embedded in a synthetic resin (Historesin or Epon 812).

In a second series of experiments, human fibroblasts were cultured on a biofoil in petri dishes. Laser-induced breakdowns were produced on the foil with laser light contucted through optical fibers. After exposure the cells underwent the same treatment as the organs. Serial semithin sections for light microscopy stained with Haematoxylin and Eosin (HE) or Methyleneblue and thin sections for electron microscopy were made both of the fibroblasts and of the the rat kidneys.

3. LASER

We used a Q-switched neodymium-doped yttrium aluminium garnet laser, usually at a wavelength of 1064 nm. More recently we also used twice and three times the initial frequency namely green light at 532 nm and ultraviolet light at 355 nm.

4. RESULTS and DISCUSSION

The physical effects of the Q-switched ND-YAG laser are described elsewhere /7,8/. Laser-induced breakdown is the intended effect of the laser pulses focussed in fluid. Material in the focus of the beam or at the point of the breakdown is transformed into the plasma state. Biologic matter is ionized and vaporized. The hot plasma expands with supersonic speed and acts as a piston on the surrounding matter. Shock waves produced by this process cause mechanical destruction of a concrement or, if they hit an organ, of cells and tissue.

We studied the effects of laser-induced breakdown and of the resulting shock wave at the surface of the kidney. The following observations were made at a laser wavelength of 1064 nm, a pulse energy of about 50 mJ and a pulse duration of 10 ns.

A macroscopic defect 200 μm to 300 μm in diameter can be seen at the point of breakdown. Semithin sections show a hemispherical to funnel-shaped crater up to 300 um deep (Fig. 1,2). The tissue adjacent to the crater wall has been destroyed mechanically to a distance of 200 μm. Vessels are interrupted, epithelial formation in renal tubules destroyed, and the basal membranes are torn open (Fig. 3,4). The content of epithelial cells is homogenized and their nuclei are deformed (Fig. 5). In some cases, the crater wall contains an electron microscopic, homogeneous zone that stains relatively strongly. Beneath it, in a dissociated cell formation, lie fragmented organelles such as destroyed mitochondria and deformed nuclei (Fig. 7,8). Few tubules are intact but their lumens are often clogged by organelle fragments.

Figure 1,2

Semithin section of Historesin embedded rat kidneys stained with H.E.

1) Crater after exposure to two laser pulses (1064 nm), 16 X

2) Crater after exposure to one laser puls (532 nm), 16 X

Figure 3,4

Semithin section of Epon embedded rat kidneys stained with Methyleneblue

3) Crater wall; dislocated nuclei, 100 X

4) Crater wall; destroyed epithelial formation in the renal tubules, 100 X

Figure 5

Semithin section of a Historesin embedded rat kidney after laser exposure stained with HE, deformed nuclei, homogenized cellls, 100 X

Figure 6

Monolayer culture of human fibroblast after exposure to one laser pulse, phase contrast picture

Figure 7,8

Thin sections of rat kidney epithelial cells near the crater wall after exposure to one laser puls.

7) deformed nucleus, 3600 X. 8) destroyed mitochondria, 12000 X

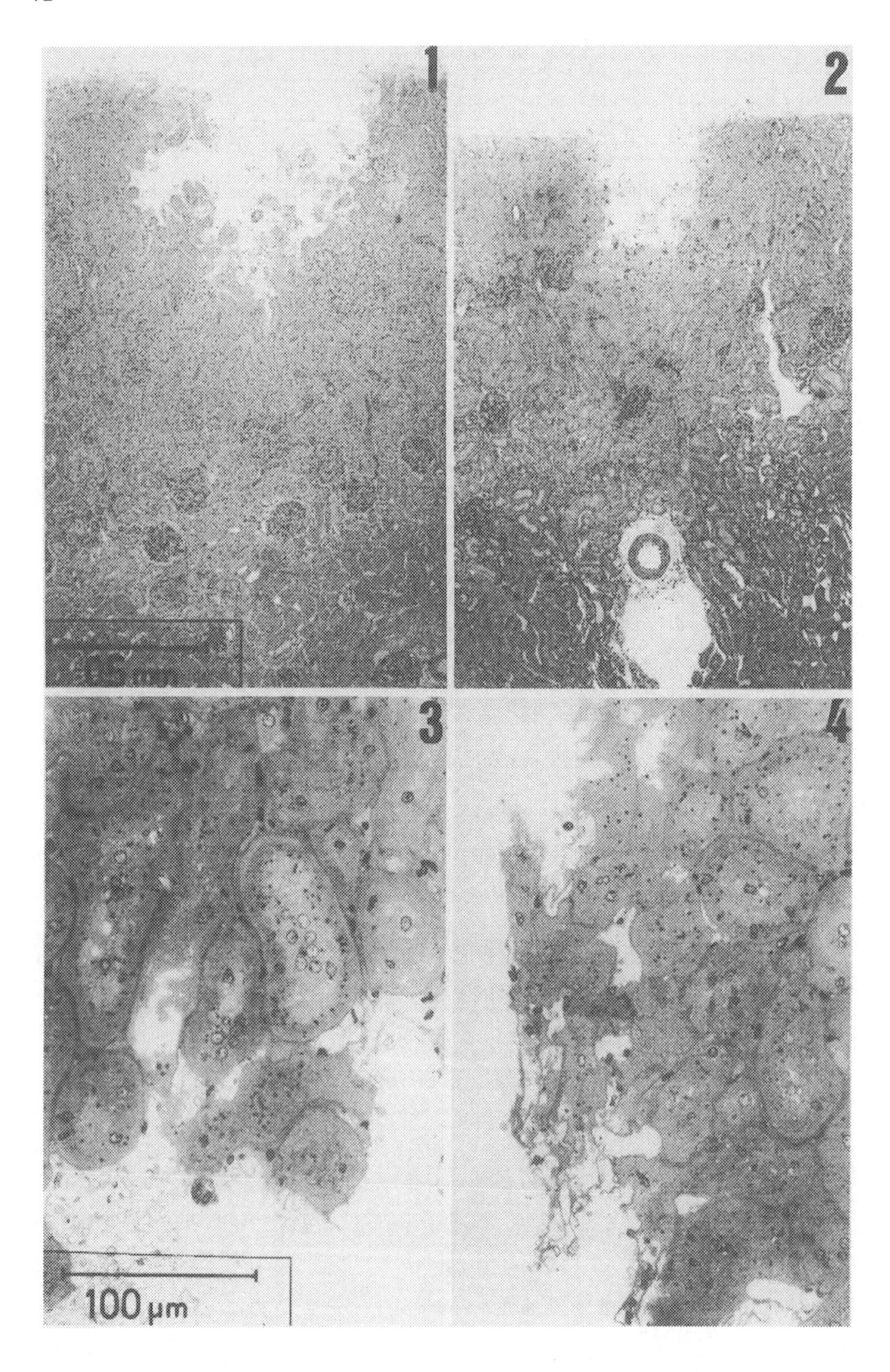
1
2
3
4
100 µm

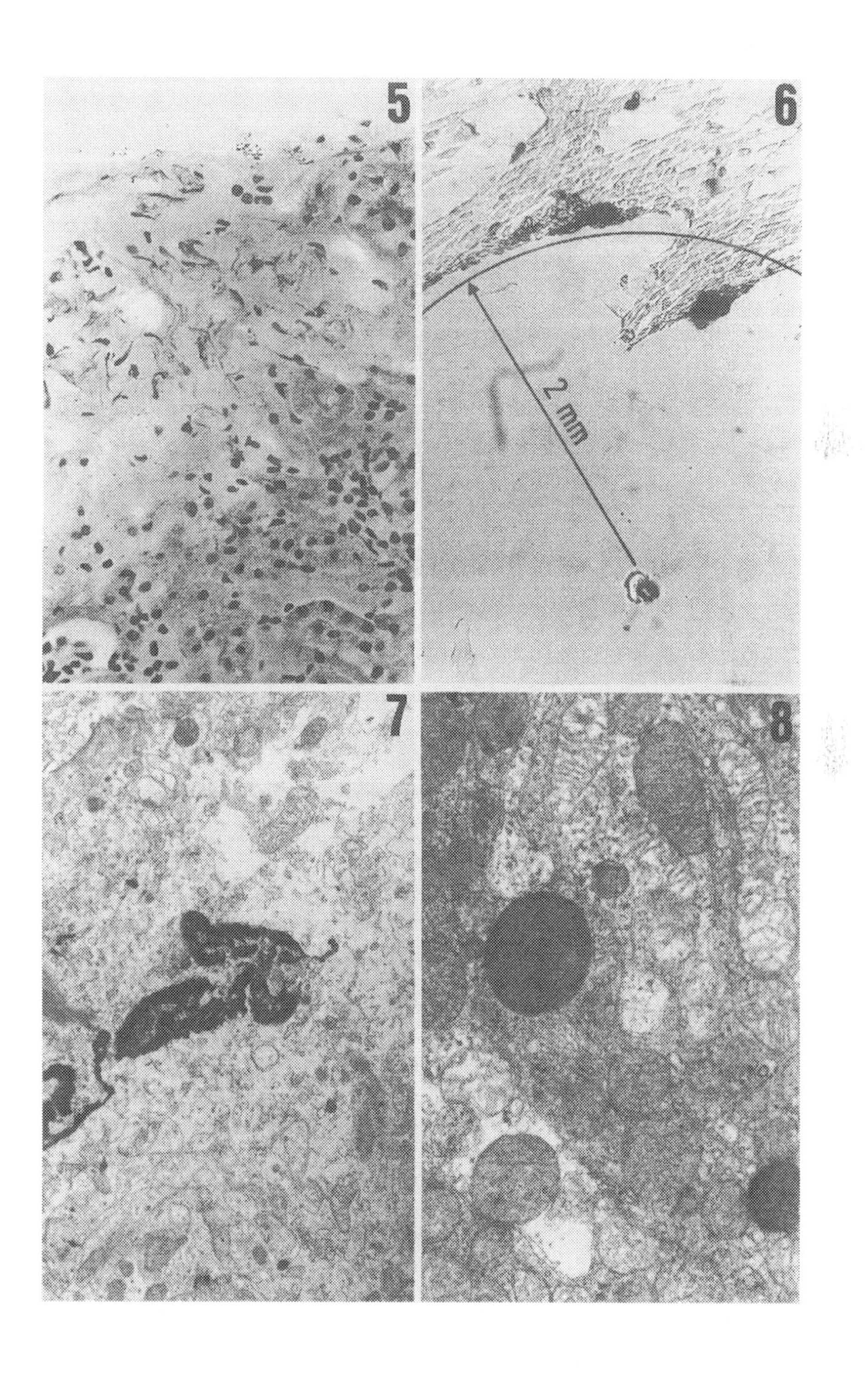
5
6
2 mm
7
8

Because of these morphologic changes, the cells in this zone that surrounds the crater cannot regenerate. The transition to non damaged tissue is gradual. Damaged mitochondria can be seen at the electron microscopic level and dislocated nuclei in some tubules at the light microscopic level. Nuclei are torn out of the cells and come to lie in the lumen of the tubules. The tissue defect itself, the crater, and the surrounding zone of mechancally destroyed tissue form a hemisphere with 500 um radius.

A comparison of the defects caused by laser at the respective wavelenghts (1064 nm - red light, 532 nm -green light and 355 nm - ultraviolet light) at constant pulse energy shows crater are of about the same diameter but a little bit smaller at shorter wave lengths. Also the walls are smoother and the crater shapes go from hemispherical to cylindrical at the shorter wavelength (532 um). The change of the cut surface from ragged to smooth with decreasing wavelength is more apparent in firmer or denser tissue such as a heart valve. But the kidney studies show that the crater-surrounding zone of mechanically destroyed tissue is smaller at shorter walvelengths. In summary, the shorter wavelengths produce craters with smoother margins and the surrounding zone of mechanical damage is smaller.

Because the laser is being evaluated for disintegration of urinary calculi and gallstones, we wanted to see how surface cells, such as the transitional epithelium of the urinary tract or the columnar epithelium of the bile ducts, react to a shock wave. As a model, we used a monolayer culture of human fibroblasts. The laser was used at a wavelength of 1064 nm and a pulse energy of 50 mJ.

The results are surprising. After the application of a single laser pulse, no intact cell could be found within 2-mm of the laser-induced breakdown. Thus the lateral propagation of the shock wave is considerable. In the next adjacent 1-mm zone, gaps can be seen in the cell formations (Fig. 6). The semithin sections show only slight morphologic changes in the individual cells. The nuclei seem shifted and some vacuoles and holes can bee seen. The results of the electron microscopic studies are in preparation.

In summary, the Q-switched Nd:YAG laser used for lithotripsy can cause injury when aimed directly at the surface of the ureter or bile duct. However, these lesions are so small that serious complications need not be expected. Using endoscopic instruments of the smallest possible diameter should help to avoid injury the epithelium of the ureter or biliary tract. The question of cellular and nuclear changes caused by the absorption of short wavelength light is still open.

5. ACKNOWLEDGEMENTS

We are grateful to Mr.R.Schmied for exellent pictures, to Mrs A.Blaschitz and Ms S.Fladl for expert histotechnical work, and to Dr.K.Tamussino for editiorial help.

This study was sponsored by the Foundation for the Promotion of Scientific Research in Austria, Project No.P 6127.

6. REFERENCES

1. H. Schmidt-Kloiber: Aktuelle Nephrologie, 1, 117-148 (1978)
2. H. Schmidt-Kloiber, E. Reichel, H. Schöffmann: Biomed. Technik, B30, Heft 7-8, 173-181 (1985)
3. R. Hofmann, R. Hartung, K. Geissdörfer, R. Ascherl, W. Erhardt, H. Schmidt-Kloiber, E. Reichel: Urologica Internation., in press (1988)
4. R. Hofmann, R. Hartung, K. Geissdörfer, R. Ascherl, W. Erhardt, H. Schmidt-Kloiber, E. Reichel: J. Urol., in press (1988)
5. E. Reichel, H. Schmidt-Kloiber, H. Schöffmann, G. Dohr, A. Eherer: Optics and Laser Technology, 19, 40-44 (1987)
6. E. Reichel, H. Schmidt-Kloiber, G. Dohr, A. Eherer, H.Schöffmann, T. Kenner: In Proceedings of the 2nd Nd:YAG-laser Conference, ed by W. Waidelich and P. Kiefhaber, Springer, Berlin-Heidelberg-New York, (1986) p.285
7. E. Reichel, H. Schmidt-Kloiber, H.Schöffmann, G. Dohr, R.Hofmann, R. Hartung: In Proceedings of the 7th Intern. Conference of the Society for Laser Surgery and Medicine, (1988), in press
8. E. Reichel, H. Schmidt-Kloiber, H.Schöffmann, G. Dohr, R.Hofmann, R. Hartung: Laser in medicine and surgery, 3, 177-183 (1987)

Analysis of Tissue Damage After Laser Induced Shockwave Lithotripsy

J. Pensel[1], *S. Thomas*[1], *P. Lieck*[1], *and G. Barreton*[2]

[1]Medizinisches Laserzentrum Lübeck, Peter-Monnik-Weg 9, D-2400 Lübeck 1, Fed. Rep. of Germany
[2]Institut für Pathologie, Medizinische Universität Lübeck, Ratzeburger Allee 160, D-2400 Lübeck 1, Fed. Rep. of Germany

Pulsed laser systems can be used to fragment in ureteral or bile duct calculi intracorporally. Experimentation with these lasers have identified the optimal parameters for lithotripsy. All systems use approximately the same pulse energy of about 30 to 80 mJ, however with different pulse lengths and wave lengths. Different transmission systems with or without special couplers produce comparable stone fragmentation into small pieces.

The advantages of the laser over alternative methods for stone fragmentation are the improved flow of irrigant permitted, the high flexibility and the reduced risk of damage to the tissue. The overall diameter of the fiber delivering the laser pulse energy is comparable or considerably smaller than that of other devices. With improvement in coupling of the laser beam with the fiber, higher pulse energies as well as high repetition rates are achieved to guarantee rapid fragmentation. For clinical use, all systems can be passed through normal endoscopes and stone desintegration is easily controlled under view.

In contrast to the other systems, the optomechanical coupler delivers a shielded plasma bubble and shock wave formation is independent from the stone or tissue. Furthermore the optomechanical coupler device combines a concentrical irrigation catheter with the laser fiber. The stone selectivity of the laser effect suggested the development of techniques that can be performed under fluoroscopic control only.

In order to demonstrate any risk of injury to the tissue caused by the laser pulse or the induced shock wave (LIB), bladder wall or the bile duct wall were brought into direct contact with the different laser probes.

We investigated the side effects of the two different laser-lithotripsy systems, the Q-switched Neodymium-YAG laser and the flashlamp excited dye laser in combination with the different probes. We tested an optical focusing system, an optomechancial coupler and the bare fiber. Between 100 and 3000 pulses at energies of

35 to 45 mJ per pulse were delivered to one precise point on the tissue. This represents the worst possible case in laser lithotripsy where the laser is discharged accidentally in direct contact with the tissue instead of the stone. The effects of the resulting injury in the tissue were studied immediately with sacrifice occurring 1 hour after the procedure and chronically with sacrifice occurring one week and four weeks later.

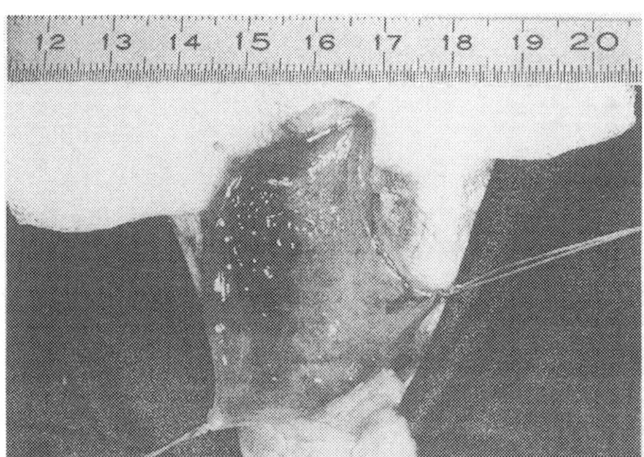

Fig. 1:
Macroscopic view of tissue injury immediately after exposition left dye laser, right Nd:YAG laser treatment

Fig. 2: Fresh bladder wall lesion after 100 dye laser pulses

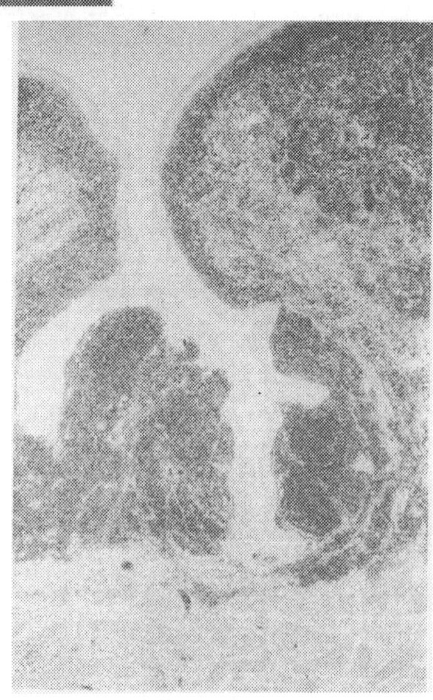

Fig. 3: Fresh bladder wall lesion after 100 Nd:YAG laser pulses

The results of the tissue reaction can be summarized:
The grade of injury seen at the tissue surface depends on the treatment modalities. Immediately after laser treatment of the inner bladder wall, the macroscopic target changings are quite different according to the laser type. Only very tiny bleeding within the mucosa can be noticed after treatment with Nd:YAG laserpulses independent of the different application systems. In contrast, macroscopic hematoma gives the impression of intensive interaction. Similar evaluation of tissue reactions were performed on the bladder wall with the optical focusing devices, the optomechanical coupler of the Nd:YAG laser and the bare fiber of the dye laser. At histology a fresh urinary bladder lesion of 100 dye laser pulses delivered from the bare and unpolished 200 µm fiber shows fresh bleeding in the lamina propria and beginning signs of necrosis. One week later granular tissue demarks the borderline of the necrosis. 4 weeks after the procedure a slight tissue fibrosis in the lamina propria and muscle layers was established. (fig. 4)

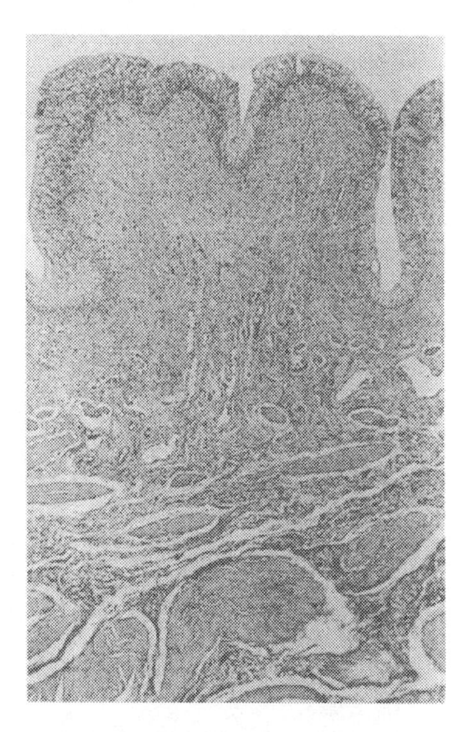

Fig. 4: Lesion 4 weeks after dye laser application

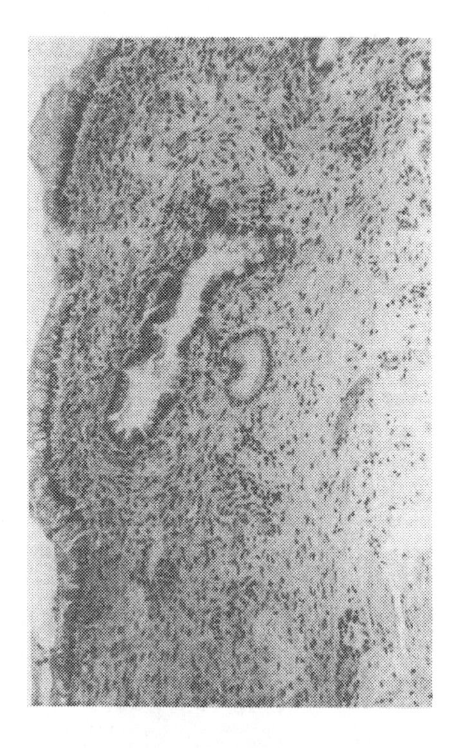

Fig. 5: Lesion 4 weeks after Nd:YAG laser application

A fresh bladder wall lesion of several 100 Nd:YAG laser pulses delivered from the optical or optomechanical device shows only superficial bleeding according to the macroscopic picture. The urothelium is removed in some areas. Four weeks later urothelium is completely regenerated and no fibrosis of the bladder wall could be found.

Scanning electron microscope evaluation of rabbit bladder treated with the dye laser shows small craterlike lesions of tissue with a diameter of 200 µm, according to the fiber core.

In addition plastic coating debris can be found on and below the epithelial surface. Pulses delivered from the opto-mechanical device lead only to superficial lesions and the mechanical contact of the probe seems to be the most traumatic part of the procedure.

The greatest potential for laser lithotripsy lies in a blind application without need of endoscopy. On account of the above mentioned side effects, the dye laser with its bare fiber has to be excluded for this modality up to now. A new development of our group for automatic regulation with a feedback mechanism makes a tissue protecting treatment possible for the dye laser too. Possibly a reduction of side

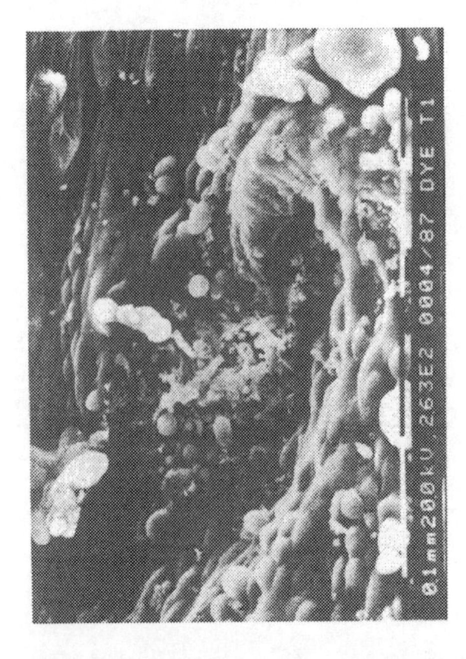

Fig. 6: Scanning electron microscope of the craterlike lesion with plastic debris

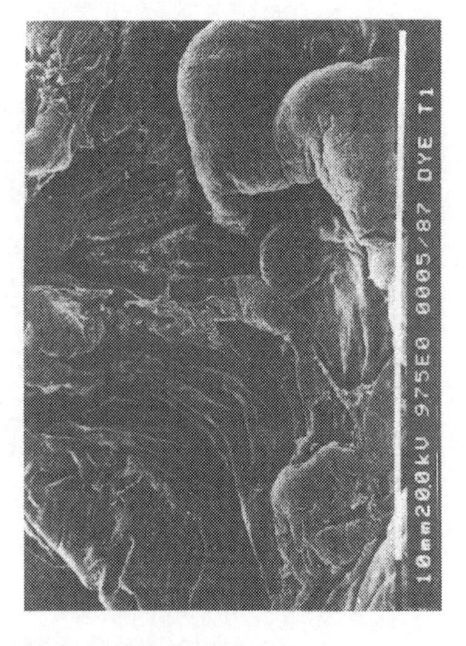

Fig. 7: The same lesion in greater magnification

effects is achievable by decreasing the pulse length and using other laser dyes, e.g. other laser wavelengths. For that reason only the Q-switched Nd:YAG laser with a miniaturized opto-mechanical coupler is used in clinical practice by us.

Tab. 1: Overview of the different tissue reactions

CLINICAL ASPECTS — TISSUE REACTIONS

coupler systems	optically focussed coupler	spherically polished fiber	optomechanical coupler	bare fiber
laser-type	Q - switched Nd : YAG			Dye
plasma bubble	free	free	shielded	free
thermal side effects	no	no	no	possible
tissue injury	minimal	minimal	negligible	substantial
scar formation	no	no	no	yes
fiber debris	no	possible (glass)	no	yes glass and plastics

MLL 87

Meanwhile, we have treated 2 patients with kidney stones, 25 with ureter stones, two with common bile duct stones and one with a pancreatic stone. Mainly in the beginning of clinical use the important problems were difficulties with the transmission system and the laser device. Due to instrumentation problems in some cases injury of the ureter wall occurred. Technical improvement has led to a simpler and more effective device. New application systems make a blind therapy possible.

1. Ell, C.: Endoscopy 18 (1986) p. 95

2. Watson, G.: J. Urol. 138 (1987) p. 199

3. Schmeller, N. T.: Laser 3 (1987) p. 184

Shock-Wave Induced Breakdown of the Tumor Microcirculation +

A.E. Goetz, R. Königsberger, P. Conzen, W. Lumper, F. Gamarra, and W. Brendel

Institute for Surgical Research, Ludwig-Maximilians-Universität, Klinikum Großhadern, Marchioninistr. 15, D-8000 München 70, Fed. Rep. of Germany

Introduction

In an intact microvascular bed the application of extracorporeal shock-waves caused transient arteriolar vasoconstriction and a concomittant reduction of blood flow. Following shock-wave exposure petechial hemorrhage, extravasation of macromolecules indicating edema formation, and intravascular thrombosis affecting predominantly the small venules were observed /2,11/. Corresponding to these findings electron micrographs have revealed ruptures of the vascular endothelium of postcapillaries and venules /2,11/.

Morphological alterations of the tumor vasculature as compared to the normal tissue vessels have already been described at the beginning of this century /14/. Therefore an insufficient functional capacity of the vascular system of tumors has been proposed /14/. Recently, few efforts have been made to exploit this knowledge for therapy of malignancies /3,4,15,16,18/. The findings after shock-wave application on intact tissue and the differences between intact tissue and the tumor vasculature led us to investigate the microcirculatory effects of shock-waves within the tumor tissue as opposed to the surrounding tumorfree areas and to study shock-wave induced effects on tumor growth and on the survival rate of tumor bearing animals.

Methods

Transparent aluminium access chambers were implanted into the dorsal skin fold of male Syrian golden hamsters /5,8,9/. 48 hours later 60 - 100 x 10^3 cells of the amelanotic hamster melanoma A-MEL-3 were implanted into these preparations. After 6 - 7 days tumors with a diameter of 6 mm were established. Details regarding relevant physiological parameters, i.e. segmental volume flow, capillary density, microvascular hematocrit and tissue oxygenation of these tumors have been reported in detail elsewhere /1,6,7,9/.

+ SUPPORTED BY KURT-KOERBER-FOUNDATION

For shock-wave exposure these tumor bearing animals were anesthetized by pentobarbital. The plastic slide covering the preparation area was carefully removed and replaced by a transparent teflone membrane. To protect the animals from drowning in the warm water-bath (37° C) during shock-wave exposure, the hamsters were placed in a water-tight plastic tube while this was continuously perfused with a mixture of oxygen and air (fraction of oxygen: 0,4). The chamber preparation with its central window area projected over this tube via a lenghtwise running slot into the water-bath.
Using the Dornier-Lithotripter XM1 the shock-waves were generated by an underwater spark discharge. In this system the discharging electrode is positioned in the focus of a rotationally symmetric ellipsoid. The shock-wave is propagated spherically, reflected by the walls of the ellipsoid and concentrated in the second focus. In this area pressures of about 80 MPa are achieved. This pressure field has a diameter of about 1,5 cm. Capacity was 90 nF and the voltage 15 kV. Prior to shock-wave exposure the chamber preparations were positioned horizontally with the intact epidermis in direction of the electrode. The window area was exactly centered in the second focus by the intersection of two laser beams /11/. The animals were randomly assigned into group 1 (control, n=7), 2 (100 shock-waves, n = 10) or 3 (2 x 100 shock-waves, 24 hours interval between the first and second shock-wave exposure; n=10). Prior to, during and after shock-waves fluorescence videomicroscopy and photomacroscopy were performed using a photoautomat and a fluorescence microscope, a high sensitivity video camera and a videorecording unit for measuring the microcirculatory changes.
For evaluating tissue perfusion a digital image processing system of the Hamamatsu Photonics Inc. (Hamamatsu Photonics Inc., Herrsching, FRG) was utilized /10/. Prior to application of the fluorochromes the video image was stored in a 16 bit video frame memory of the image processor. Following intravenous administration of the fluorochromes FITC-Dextran 150 and sodium fluorescein a new video image was transferred to a second video frame memory. Subsequently the fluorchrome image was subtracted from the baseline image. Accumulation of the fluorochromes was indicated by a signal intensity increase in the subtracted image. Measurements were carried out repeatedly in 5 min intervals over a period of 30 min. Thus even if the vessels were obscured by hemorrhage, perfusion of the tissue could be monitored by visualizing the diffusion into the hemorrhage.

Results

In the control group no significant changes of the microcirculatory parameters, such as blood cell velocity, vessel diameters, or segmental volume flow were measured when keeping the animals in the warm water-bath for a period necessary for shock-wave treatment.

In the surrounding tumor free tissue of these tumor bearing animals application of 1 x 100 shock-waves led to the constriction of all arteriolar microvascular segments, i.e. of all arcading, transverse and terminal arterioles and of precapillaries. Within these tumor free areas arteriolar vasoconstriction started within seconds, reached its maximum after about 20 - 30 s and tended to release after 4 - 10 min. In addition, microhemorrhages were detected at the site of small venular microvascular segments. In these areas leakage of macromolecules and the appearance of amorphous conglomerates were observed. Within the tumor areas complete hemorrhage after application of 100 shock-waves was noted. Planimetric analysis indicated significantly elevated hemorrhage in the tumor tissue. 50% of the adjacent tissue area was hemorrhaged as opposed to 100% in the tumor.

Accumulation of the sodium fluorescent dyes FITC-dextran 150 and sodium fluorescein could not be demonstrated in the tumor at 30 min after shock-wave treatment. 12 hours later reperfusion began at the tumor edges in 50 % of the tumors and at 24 hours after shock-wave treatment all tumors were reperfused. However, we were unable to differentiate if reperfusion occured in previously occluded vascular segments or in blood vessels formed by neoangiogenesis. Due to the complete hemorrhage of the tumor, no quantitative analysis of the microcirculatory blood flow within the tumor area after reperfusion was possible. We were only able to assess if perfusion was present or not.

Addition of further 100 shock-waves induced an increase of hemorrhagic areas in the surrounding tumor free tissue. Perfusion in the surrounding tumor free tissue after the second administration of 100 shock-waves was noted about 10 minutes after shock-wave application. Within the tumor the collapse of microcirculation lasted for more than 24 hours (Fig. 1).

Tumor growth curves in the control group and in the animals treated with shock-waves are shown in Fig. 2. 15 days after tumor cell implantation the chamber preparation in the control group was completely covered by the amelanotic hamster melanoma. Tumor growth in the animals treated with 100 shock-waves was significantly delayed 3 days after shock-wave treatment, but did not differ significantly from the controls 5 days later. In group 3, the hemorrhagic area 48 hours after the second 100 shock-wave application was significantly smaller than the tumor mass in the control group. 5 days later an apparent reduction of the hemorrhagic mass was observed. The newly formed blood vessels at the edges of these hemorrhagic areas did not show the typical chaotic angioarchitecture of the amelanotic hamster melanoma.

As to the survival rates the following findings were observed: Maximum survival in the control group was 32 days and in the group treated with 1 x 100 shock-waves 63 days, whereas in group 3 70% of the animals were living for more than 9 months without any recurrence of the tumor. The hamsters of this group were sacrficed at that time. 30% of the animals in group 3 died either due to a hemorrhagic shock after removal of the catheters by the animals themselves (n=2) or from a rectum prolapse (n=1). In the control group and in group 2 (100 shock-waves) 6 out of 7 animals and 9 out of 10 hamsters died from tumor cachexia and multiple metastasis formation, respectively.

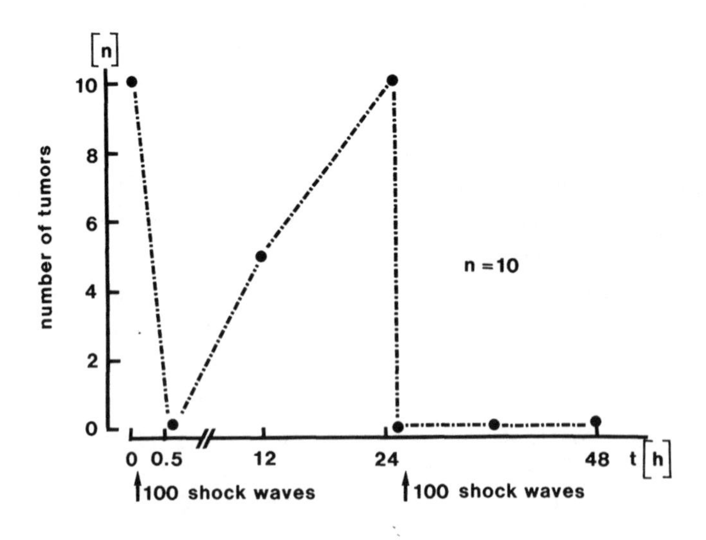

Fig. 1. Number of perfused tumors as evaluated by videofluorescence microscopy and digital image processing at 30 min, 12 h and 24 h after initial application of 100 shock waves and at 30 min, 12 h and 24 h after application of additional 100 shock waves

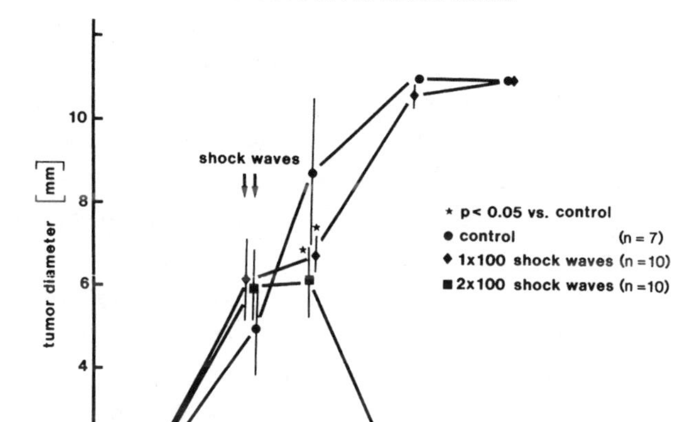

Fig. 2. Tumor growth curve in control and in shock wave treated animals. Circles, controls (n=7); rhombi, 1 x 100 shock waves (n=10); squares, 2 x 100 shock waves at an interval of 24 h (n=10). Mean values and standard deviations (vertical lines) are shown; asteriks indicate p . 0,05 versus control

Discussion

The present findings of a complete hemorrhage in the tumor area as opposed to a partial hemorrhage in the surrounding tumor free tissue and of a persisting breakdown within the tumor tissue demonstrate that the vasculature of the tumors is more sensitive to shock-wave treatment as the surrounding tumor free tissue or the skin muscle microcirculation of the Syrian golden hamsters.

The most intriguing question therefore to be answered in this context is which mechanisms are responsible for inducing a longlasting circulatory arrest of the tumor microcirculation. In this context the specific characteristics of the vascular supply of the tumors have to be considered. The characteristics of the microcirculation of the amelantic melanoma A-Mel-3 accord with those of other different experimental tumors /1,3,6,7,9,12,13,14,17/.

Tumor vessels are dilated, tortuos and exhibit redundant curvatures. Wide sinusoids may be found as well as a critical reduction of the capillary density, being insufficient for an adequate supply of the tissue. In addition it has been shown that blood flow is extremely nonhomogeneous since high and low flow perfusion areas as well as shunt perfusion may be found simultaneously. Moreover it has been demonstrated that the vascular resistance in the tumor microvessels is increased due to these specific characteristics and due to the most important characteristics i.e. "the insufficient secondary neoangiogenesis". Electronmicrosciopic studies revealed that mesenchymal cells participate at the lining of the vascular wall and endothelial gaps can be seen allowing extravasation of intravascular cells and macromolecules. Moreover endothelial edema is sometimes present as well. Therefore intravascular thrombi formation is seen in-vivo and moreover an enhanced microhematocrit is found in the tumor vessels being 100% above the level of the normal tissue /9/. These factors contribute to an enhanced rigidity of red blood cells and thereby to an elevated vascular resistance of the tumor microcirculation. It seems therefore plausible that therapeutic induction of a breakdown in the tumor microcirculation may be accomplished more easily due to its state of circulatory dysfunction relative to normal tissue /3/. In addition shock waves induce arteriolar vasoconstriction in the surrounding tumor free tissue and formation of intravascular thrombosis of the tumor vessels, as well as selective hemorrhage in the tumor areas. These additional factors contribute to a further increase of the vascular resistance of the tumor microcirculation and could lead consequently to complete ischemia within the tumor areas. A vicious cycle is probably activated leading to no reflow in the tumor and eventually to a self-destruction particularly of the tumor center.
So far our conclusions are, however, restricted to the the skin fold chamber preparation. Studies are therefore needed on solid tumor models to test if a tumor treatment may be performed by shock-wave application only.

References

1. K. Ashiashi, B. Endrich, A. Goetz, K. Messmer: Canc. Res. 41 , 1898 (1981)
2. W. Brendel, M. Delius, A. Goetz: Progr. Appl. Microcirc. 12 , 41 (1987)
3. J. Denekamp: Progr. Appl. Microcirc. 4 , 28 (1984)
4. H.A. Eddy: Radiology 137 , 512 (1980)

5. B. Endrich, K. Asiashi , A. Goetz ,K. Messmer: Res. exp. Med. 177 , 125 (1980)
6. B. Endrich , A. Goetz, K. Messmer: Int. J. Microcirc. Clin. Exp. 1 , 81 (1982)
7. B. Endrich, F. Hammersen, A. Goetz, K. Messmer: J. Natl. Cancer Inst. 68 , 475 (1982)
8. A. Goetz, B. Endrich, C. Laprell, K. Messmer: Bibl. Anat. 20 , 65 (1981)
9. A.E. Goetz: MD Thesis, Ludwig-Maximilians-Universität München (1987)
10. A. Goetz, J. Feyh, H. Ortner, P. Conzen, W. Brendel: Int. J. Microcirc. Clin. Exp. 6 , 71 (1987)
11. A.E. Goetz, R. Königsberger, F. Hammersen, P. Conzen, M. Delius, W. Brendel: This volume
12. F. Hammersen, U. Osterkamp-Baust, B. Endrich: Mikrozirk. Forsch. Klin. 2 , 15 (1983)
13. F. Hammersen, B. Endrich, K. Messmer: Int. J. Microcirc. Clin. Exp. 4 , 31 (1985)
14. H. Ribbert: Dtsch. Med. Wochenschr. 30 , 801 (1904)
15. C. Song: NCI Monogr. 61 , 169 (1982)
16. W. M. Star, P. A. Marijnissen, A. E. van den Berg-Blok, J. A. C. Versteeg, K.A.P. Franken, H. S. Reinhold: Canc. Res. 46 , 2532 (1986)
17. B. A. Warren: In Tumor blood circulation. ed. by H. I. Peterson (CRC Press, Boca Raton, 1979) p. 49
18. A. Young: Br. med. J. 283 , 114 (1981)

Limits of Optical Fiber Systems for Pulsed Lasers

P. Hering

Max-Planck-Institut für Quantenoptik, D-8046 Garching, Fed. Rep. of Germany

We have used common pulsed lasers to determine the transmission characteristics of different step-index quartz fibers with core diameters of 200 μm up to 1500 μm. The pulsewidth varied from 10 ns up to 1.5 μs and the applied energies up to 0.5 joules per pulse.
In general, the maximum transportable energy through fibers is limited by laser induced damage at the surface rather than bulk damage. Thus it is very important to reduce intensity spikes in the beam profile.
In the UV and in the infrared region strong absorption reduces the application of fibers in addition to surface damage whereas nonlinear processes like stimulated Raman or Brillouin scattering have minor influence, particularly for short fiber lengths.

Introduction:

The development of new powerful pulsed lasers has increased the number of applications in the medical and industrial field. For reasons of practicability and safety the use of optical fibers is often necessary to bring the laser light from the laser itself to the point of action. A fiber length of a few meters is usually sufficient.

The problems involved in the application of pulsed lasers to optical fibers are the following:

1. Laser induced damage (LID):
The energy density (J/cm^2) or/and the power density (W/cm^2) of the applied laser exceeds the damage threshold of the fiber material either on the surface or bulk of the fiber. This is the most severe limiting factor and can only be overcome by using large core diameter fibers.

2. Linear and nonlinear processes:
The transmission through the fiber is limited by linear and

nonlinear absorption, Rayleigh scattering and stimulated Raman and Brillouin scattering. Particularly in the UV and IR (e.g. CO_2 Laser wavelength) absorption dominates the losses and thus limits the application.

Experimental:

In our experiments we only used standard fibers, fiber holders, lenses etc. Thus our results are easily transferable to other laser labs. The results we are presenting in this paper are average values which can easily be obtained.

We applied different pulsed lasers available in our institute. The standard characterization of these laser as pulse energy and pulse width are obtained with usual energy meters and fast photodiodes. We were not able to characterize these lasers always in terms of spacial and temporal intensity beam profile or mode pattern which would be necessary to make our results fully reproducable to others.

Fibers and fiber conection:

We only used step index fibers with quartz core. The large core diameter fibers were plastic clad silica (PCS) fibers from Quartz et Silice, Paris or Fiberguide, Stiling N.Y., both special UV grade. For the 200 µm core diameter fiber we choosed AS (all silica) fibers with core and cladding made of quartz.

Originally we used commercial fiber connectors with modified bore diameter. The fiber tip was glued into the connector and polished to optical quality. Using this setup it was necessary to prevent the laser beam hitting the metal part of the connector because optical breakdown on the metal surface would cause subsequent optical degradation of the whole connector. Now we polish the bare fiber and mount it protruding the fiber holder roughly 20 mm. With this kind of mounting the positioning of the fiber gets uncritical particularely when the focus diameter is larger than the core diameter. This is e.g. always the case with focus diameters of flashlamp pumped dye lasers coupled into a 200 μm fiber.

A given laser beam diameter has to be adjusted to the core diameter of the fiber. Generally pulsed lasers deliver much more energy per pulse and cm^2 than fibers can survive. We used standard convex lenses with focal length between 100 to 200 mm. The fiber tip is placed at that point where the laser beam diameter and core diameter coincide. With short focal length lenses one has to be aware of optical breakdown in front of the fiber entrance surface. The experimental setup is rather simple and is described elsewere /1/. We measured the laser pulse energy which is coupled into the fiber (not the total energy of the laser) and the enèrgy transmitted through the fiber and determined the transmission and attenuation. We increased the laser energy by removing neutral density filters until the laser damage threshold was reached. In order to get a feeling where to expect the damage threshold we first used a simple quartz plate for detection. Coupling losses due to reflection at entrance and exit surface are included in the transmission data but are not important. In some cases the losses due to absorption inside the fiber were not measurable with different fiber lengths.

Transmission and attenuation:

Table I gives an overview of the applied lasers with laser and fiber data. This table gives an orientation about the possibilities of the transmission of pulsed lasers through fibers. It should be mentioned that these data are not the highest possible values but standard ones. In particular the transmission values of our Nd:YAG laser are low due to spiking problems. We would like to highlight the transmission data of 2 dB/m for the ArF laser at 193 nm which are as far as we know reported here the first. Although the transmission is still rather poor it is an enormous progress compared to earlier attempts with attenuations of 100 dB/m.

Table I:

Transmission and attenuation for different pulsed lasers through optical fibers

Wave-length (nm)	Laser	Pulse width (ns)	Fiber-length (m)	Fiber-core (µm)	Trans-mission	Attenu-ation (dB/m)	Energy Input (mJ)
193	ArF	15	1	1000[1]	10 %	10	4
			1	600[1]	10 %	10	1
248	KrF	15	1	1000	50 %	3	4
308	XeCl	15	1.6	1000	74 %	1.3	30
351	XeF	15	1	1000	78 %	1	40
			1	600	78 %	1	19
570-610	Dye Flash-lamp	1500	2	1500	84 %	<0.4	500*
			8	1000	84 %	<0.1	500*
			1	200[2]	84 %	<0.8	130
730		2000	2	200[2]	90 %	<0.3	150 [4]
450-610	Dye Excim.	12	2	1000	90 %	<0.3	40
			8	1000	90 %	<0.06	40
694	Ruby	30	2	700[3]	40 %	2	80 [5]
1064	ND-YAG	20	2	1000	95 %	<0.1	55
533	"		2	1000	94 %	<0.1	25
355	"		2	1000	91 %	<0.2	10
266	"		2	1000	50 %	0.5	4*

1500 µm fiber: PCS Quartz et Silice

1000 µm fiber: PCS Q. et S.

600 µm fiber: PCS Q. et S.

[3] 700 µm fiber: Schott

[2] 200 µm fiber: AS Q. et S. (all silica)

[1] 1000µm fiber: Fiberguide, special UV fiber

[1] 600µm fiber: Fiberguide, special UV fiber

[4] R.Engelhard, Medical Laser Center Lübeck, private commun.

[5] A.Ettemeyer, G.Urmann, Rottenkolber, Kirchheim, priv. commun.

* applied laser energy was not high enough to reach damage threshold

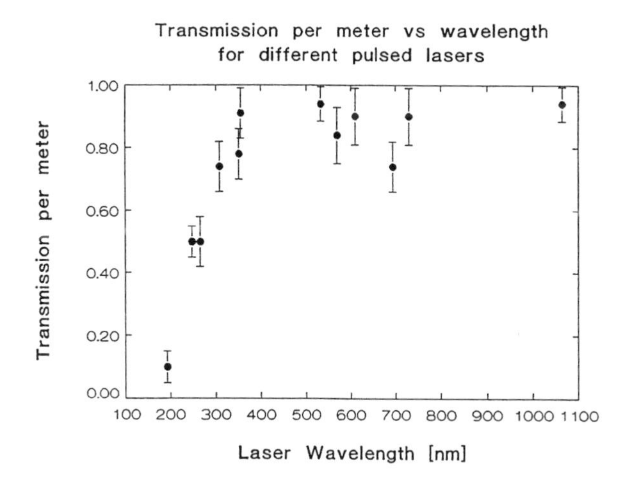

Figure 1 displays the wavelength dependence of the transmission through different fibers which sharply drops at 300 nm due to absorption inside the fiber. We could not detect any nonlinear conversion effects like Raman or Brillouin scattering because of the short fiber length we used. Such processes are however reported in the literature and must be taken into account for longer fiber lengths /2/.

Laser induced damage:

In the visible the transmission through a fiber stays nearly constant for increasing laser energy until the fiber entrance surface is suddenly destroyed. Decreasing transmission for higher laser energy in the UV is reported /3/ as well as decreasing transmission due to long irradiation with UV laser light and subsequent color center formation /4/.

The process of laser induced damage is studied intensively but nevertheless not very well understood. Many processes are involved with changing partion. We have to refer to the literature at this point and recommend the Boulder Damage Conference Reports edited by the National Bureau of Standard (NBS).

The catastrophic degradation of the fiber surface has different origins:

1. intrinsic linear absorption
2. impurity and inclusion absorption
3. multiphoton absorption
4. self focusing

All these mechanisms can cause optical breakdown close to or on the fiber surface followed by an intensive acoustic shock wave damaging the fiber. Table II summerizes the involved effect with its intensity thresholds.

Table II:

Mechanism for Laser Induced Breakdown		
critical electron density for breakdown:		$>10^{18}$ cm^{-3}
mechanism involved:	1. Self focusing	$>10^{4}$ W/cm^{2}
	2. Multiphoton ionisation	$>10^{6}$ W/cm^{2}
	3. Avalanche process	$>10^{6}$ W/cm^{2}
	4. Field ionisation	$>10^{9}$ W/cm^{2}
		10^{10} W/cm^{2} = $2*10^{6}$ Volt/cm

The laser induced damage threshold depends on wavelength, pulsewidth, energy density (sometimes called fluence) (J/cm^{2}) and irradiance (sometimes also called intensity) (W/cm^{2}) and thus strongly on the laser beam characteristics.

To compare results one has to know precisely all laser parameters including the mode pattern, polarisation, two dimensional intensity profile and time behavior of the laser pulse. Glas processing characteristics, impurities, polishing history, surface condition and environmental situation differ in general. Thus general scaling laws cannot be provided.

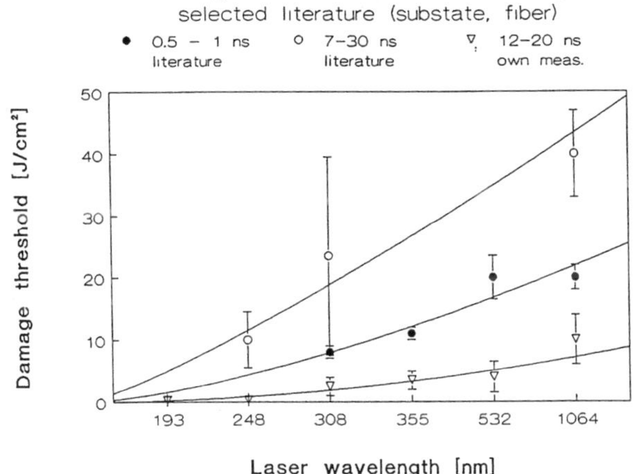

Figure 2 shows the wavelength dependence of the laser induced damage threshold. We took literature values /5/ of selected laser wavelengths and two pulse length ranges for quartz material either as substrat or as fiber. This figure is rather confusing but displays the problematic in this field very well. The error bar shown in fig 2 are not the result from a single measurement but display the results from different authors. In particular at 308 nm the variation is rather unusual and the high values of 40 J/cm^2 are doubtable /6/.

All three sets of data show the same trend for the wavelength dependence. Our data values are low compared to other literature values but were conservativly stated and were obtained differently with simple polishing methods and without exact evaluation of our intensity beam profile. Thus higher values could probably be obtained with more effort in preparing the fiber entrance and exit surfaces.

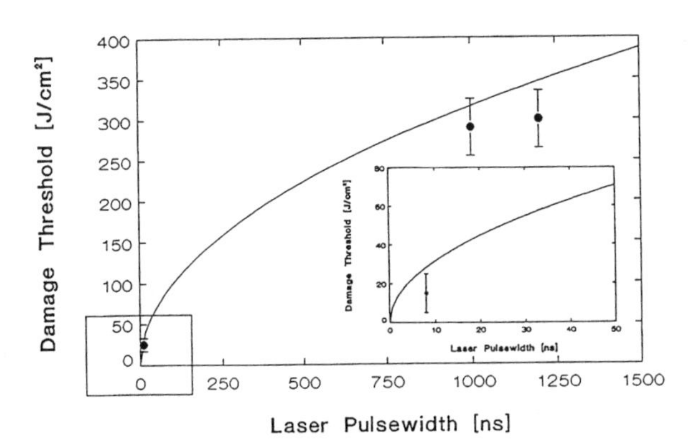

The damage threshold dependence is expected to scale with $t^{1/2}$ /7/. Fig. 3 shows this behavior for one specific laser wavelength at 590 nm. This result is of special interest for the application in laser lithotripsie where the Nd:YAG laser with approx. 10 ns pulse width and the flashlamp pumped dye laser with 1.5 µs is mainly used /8/. Table II gives a comparison of the lasers with the corresponding fiber diameters.

Table III:

Comparison Nd:YAG - Dyelaser

Nd:YAG	µm	GW/cm²	J/cm²
50 mJ	200	15.9#	159.2#
10 ns	600	1.8	17.9
Dye laser	µm	MW/cm²	J/cm²
50 mJ	200	106.2	159.2
1.5 µs	600	11.6	17.2

\# above laser damage threshold

The advantage of the long pulse dye laser is obvious. To deliver the same energy to the surface of a kidneys stone a 200 µm fiber can be used instead of a 600 µm fiber. Such a fiber is much less stressed because the damage threshold is more than 10 times larger than it is for the short pulses of a Nd:YAG laser.

Conclusion:

Table IV summerizes our results. Better fiber quality and better surface treatment will certainly improve the damage threshold in the future in particular the use of tapered fibers in the UV.

Table IV:

Conclusion
Transmission limited by:
a: laser induces surface damage b: UV and IR absorption
nonlinear effect not important for short fibers
damage threshold increases with laser pulse duration ($\propto t^{1/2}$)
important: good surface quality! smooth laser beam profile (measurement necessary!)

Literatur:

1. L. Prause, P. Hering: Laser und Optoelektonik Nr.1/1987 p.25
2. M. Rothschild, H. Abad: Opt. Lett. 8, 653 (1983)
3. Y. Itoh, K. Kunitomo, M. Obara, T. Fujioka: J. Appl. Phys. 54, 2956 (1983)
4. E.A. Nevis: SPIE Vol.540 Southw. Conf. on Optics, p.421 (1985)
5. see literature cited in 1.
6. D.L. Singleton, G. Paraskevopoulos, R.S. Taylor, L.A.J. Higginson: IEEE J. Quant. Electr. QE-23, 1772 (1987)
7. R.S. Taylor, K.E.Leopold, S.Mihailov: Opt. Comm. 63, 26 (1987)
8. see contributions in this volume

Fig. 1: wavelength dependence of fiber transmission

Fig. 2: wavelength dependence of laser induced damage threshold

Fig. 3: damage threshold dependence of laser pulse width

Enhanced Energy Transport via Quartz Fibers Using 250 ns Long-Pulse Excimer Laser

U. Sowada, H.-J. Kahlert, W. Mückenheim, and D. Basting
Lambda Physik Forschungs- und Entwicklungsgesellschaft mbH,
Hans-Böckler-Straße 12, D-3400 Göttingen, Fed. Rep. of Germany

Experiments have shown that a significantly enhanced energy transport via quartz fibers is achieved by using a long-pulse excimer laser. The technology used to generate 250 ns excimer laser pulses is described. Additionally it is reported on considerations leading to the design of an excimer laser fiber coupler. Measured data are presented.

Long-Pulse-Excimer Laser Technology

In comparison to a standard excimer laser the long-pulse excimer laser has been designed to operate at 308 nm using the gas mixture XeCl. At a maximum repetition rate of 100 Hz the energy output per pulse is 150 mJ. The optical resonator forces the laser to run in stable cavity modes. This ensures a beam size of 15 mm x 15 mm. The horizontal and vertical divergence of the beam emitted is about 5 mrad. Depending on the discharge circuit used the achieved pulse-length is about 250 to 300 ns. To achieve the long pulse a highly reliable solid state pulse forming network (PFN) has been designed. A DC-power supply is used to charge the PFN. The excimer gas is preionized by a separate prepulse circuit which is used simultaneously to trigger the main discharge. This prepulse power is switched by a thyratron.

Fiber Coupling

In optical fibers the effect of total reflection is utilized for light guiding. Optical fibers normally consist of an arrangement of 3 materials:

- a cylindrical quartz core (typical diameter 100 microns to 1000 microns)
- a cladding (doped quartz or highly UV-transparent polymer)
- a mantle (distributes bending forces, source of flexibility)

The damage threshold of optical materials depends on power density rather than energy density. Damage can occur in the core and in the cladding. The typical power threshold is about several GW per cm^2 using the wavelength 308 nm. Due to the angle of deflection, the power density of the pulsed laser light at the cladding surface is much smaller than at the fiber face. But this is only valid if the fiber is exposed to light rays which fit into its acceptance cone. Physically the optimum pulse energy transmission is achieved if all modes of the fiber are filled evenly. The best method is to image a circular aperture which is irradiated on one side by the laser beam. A demagnification ratio up to ten has produced reliable results. Using a long-focal-length lens it is possible to fill a fiber up to 80%. Using optimized optical fiber coupling systems with respect to the special excimer beam properties a standard EMG 102 MSC and the long-pulse laser have been used for fiber coupling experiments at the wavelength 308 nm. We have observed that fiber aging occurs at high energy transmittance and also high repetition rates. Using the 20 ns pulse of a standard excimer laser at 100 Hz repetition rate, 6 mJ can be transmitted through a 600 micron fiber reliably. Using the 250 ns pulse, the energy output of a fiber of 1 m length is 40 mJ.

Ablation Tests

First investigations have been performed to measure the threshold energy density for ablation using the long-pulse excimer laser. In case of polyimid (Kapton) it is well known that ablation occurs, using a 20 ns standard excimer laser, at energy densities of about 50 mJ per cm^2. The following data have been measured.

Table 1:

Kapton	ablation threshold
20 ns	(50 +- 10) mJ/cm^2
250 ns	(70 +- 10) mJ/cm^2

The only slightly higher threshold energy density for the 250 ns pulse shows the expected behaviour. Ablation of non-optical materials like polyimide mainly depends on the energy density. In vitro experiments have shown effective gall-stone destruction is possible with about 150 pulses of 40 mJ. This test has been performed with 308 nm long-pulse excimer laser light delivered via a 600 micron quartz fiber.

Conclusions

For all ablation applications it is important that the ablation threshold energy density is nearly independent of the pulse length. Using the 250 ns pulse of a long-pulse excimer laser up to 50 mJ per pulse can be transmitted through a quartz fiber. This pulse energy is sufficient to crack and pulverize gall-stones. However, the reported data are only preliminary and much more systematic work is necessary.

Ray-Tracing Calculations of the Focusing Efficiency of Spherical Fiber Ends*

Ch. Hauger, F.-W. Oertmann, and W.-G. Wrobel

Aesculap-Werke AG, D-7200 Tuttlingen, Fed. Rep. of Germany

1. Introduction

Optical fibers with spherically shaped tips have been proposed and used in surgery for different applications such as laser angiosurgery /1/, photodynamic therapy /2/ and laser lithotripsy /3/. In the latter case there are two reasons for not using plane fiber ends:

a) A plane fiber end emits divergent radiation. Therefore the laser-induced breakdown takes place exactly at the fiber end, thereby damaging the fiber tip. By shifting the point of highest intensity from the fiber end, there is less probability of damage.

b) If the breakdown threshold energy can be reduced by focusing, stone destruction can be achieved with less laser energy, which means less danger of thermal tissue damage.

With our ray-tracing program we investigated the focusing efficiency of spherical step-index quartz fibers immersed in water.

2. Calculation Method

We calculated the intensity distribution in the image space at several planes vertical to the direction of the optical axis. The surface curvature of the fiber tip and the fiber numerical aperture were varied.

In most calculations we assumed a PCS 600 step-index quartz fiber with 600 micron core diameter. Therefore the intensity distribution at a fiber cross section is assumed to be homogeneous. Since transmission by multimode fibers destroys optical coherence, it is sufficient to evaluate light intensities only.

---

*) Work supported in part by Bundesministerium für Forschung und Technologie

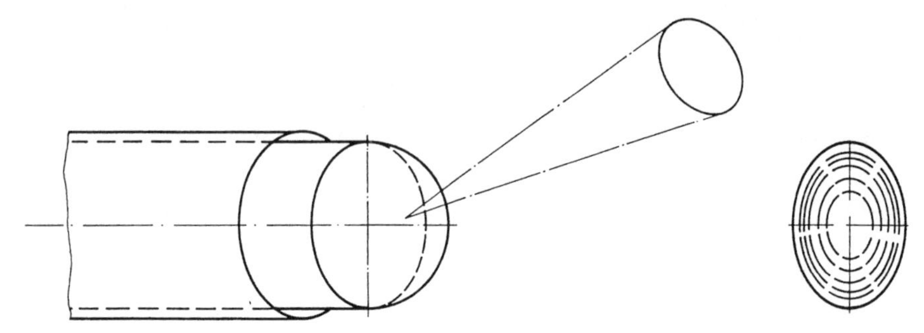

Fig. 1 Ray-tracing method for calculating the intensity enhancement of spherical fiber ends

Every point at the illuminated fiber end emits ray pencils (Fig. 1), whose aperture angle is determined by the fiber's numerical aperture. The intensity distribution at any diameter vertical to the cone is also homogeneous.

As every calculated ray cone represents an equal-size area of the fiber cross section we arrange them in equal-size area circular rings as indicated in Fig. 1. This arrangement is also valid for the single rays of the proper ray cone. With the single calculated rays being arranged in this way, they represent the same beam energy in the image plane, which simplifies the evaluation considerably.

At the fiber end a plane/convex lens is attached. For the calculation it does not matter whether a separate lens is put at the fiber end or whether the fiber end itself is melted or polished spherically.

In the image plane we can choose a random number of image fields arranged equidistantly and vertically to the optical axis. These image planes are divided in circular rings in the same way as the fiber end.

We then determine the ray vector in the image space. With these numbers we can easily calculate the ray intersection points at several image planes. It is not necessary to evaluate the whole ray bundle for every defocussed image.

3. Results

First we investigated the dependence of the focusing efficiency on the lens radius. In the following diagrams the numerical aperture of the fiber is NA = 0.2. The ratio d (fiber half diameter : lens surface radius) is varied. d = 0 corresponds to a plane fiber end, and d = 1 is a hemispherical fiber tip. In Fig. 2 to 5 the distance between the respective image point and the fiber end is depicted at the abscissa (in Fig. 2 the gaussian image plane is situated outside the diagram). The ordinate is the percent energy contained in image circles with radii of 0.05, 0.1, 0.15, 0.2, 0.25 and 0.3 mm.

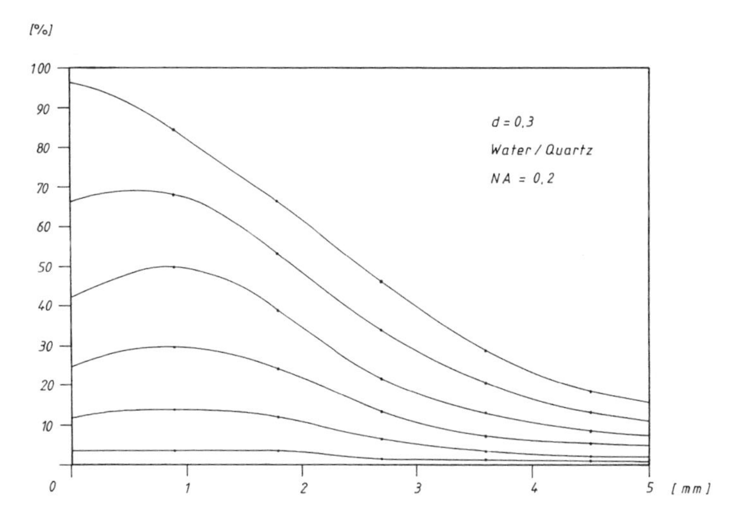

Fig. 2 Intensity enhancement of a moderately shaped spherical fiber end

In Fig. 2 a clear maximum is not observable. Obviously with d = 0.3 there is negligible focusing. The situation improves with higher d values. Fig. 3 shows data for d = 0.9. The intensity maxima are much more pronounced and are very close to the fiber end - much closer than the gaussian image plane. However the intensity enhancement (the energy contained in a circle at an image plane compared to the value at the fiber exit plane) is less than 2.

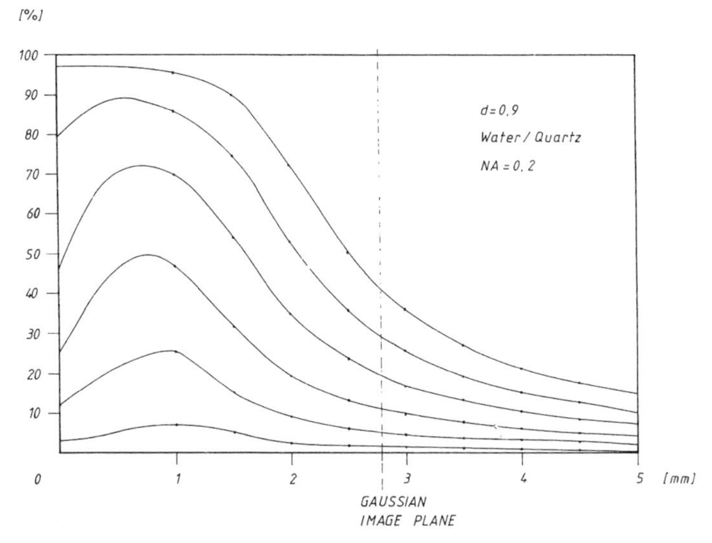

Fig. 3 Intensity enhancement of a nearly hemispherical fiber tip

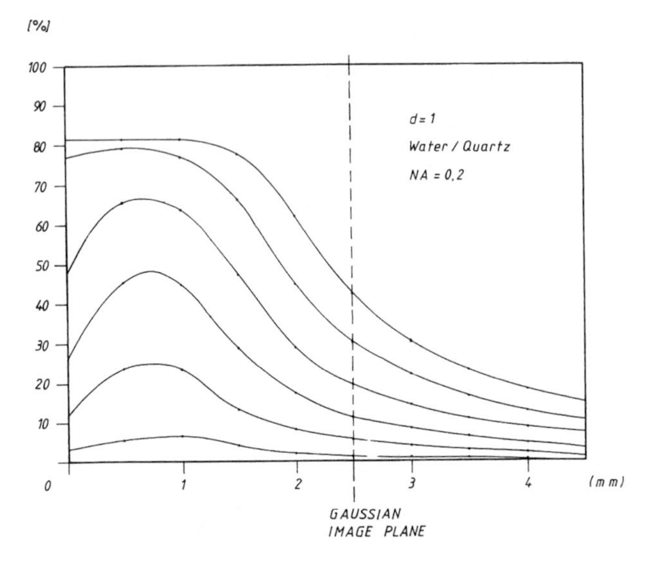

Fig. 4 Intensity enhancement of a hemispherical fiber tip

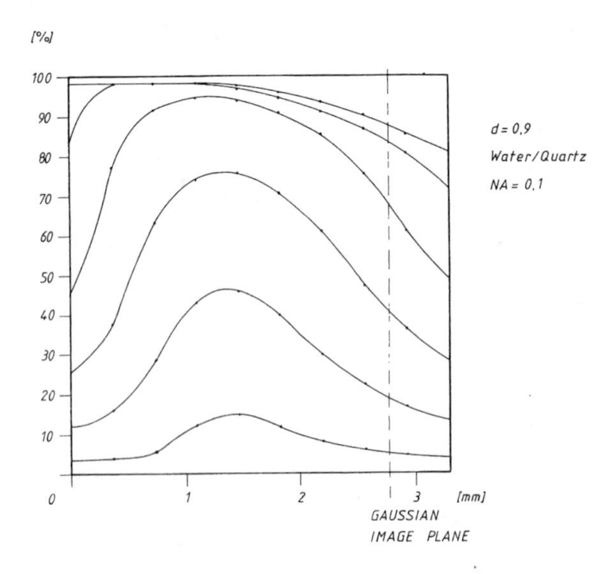

Fig. 5 Intensity enhancement of a near hemispherical fiber tip at a low-aperture fiber end

With d = 1 (Fig. 4) there is only a gradual change. However total transmission is limited to 82 %, mainly due to total internal reflection. The most striking result is that nearly independently of d the point of maximum intensity enhancement is always situated about 0.8 to 1 mm in front of the fiber end. This is only about 1.5 times the core diameter - far too close to the fiber end to avoid damage.

Some improvement can be expected from fibers with lower numerical aperture (NA = 0.1 in Fig. 5). The point of maximum intensity enhancement shifts to the gaussian image plane, and the value of the intensity enhancement increases considerably to about 3 in a distance of 1 to 2 mm.

Probably fibers with higher index cores or aspheric fiber tips might still improve the situation.

4. Conclusions

Standard PCS fibers with spherical fiber tips are inefficient focusing elements. The use of low aperture fibers is essential for good intensity enhancement.

References

1. C. J. White et. al.: Lasers in Surgery & Medicine, 7 (1987) 81
   H. J. Geschwind et. al.: ibid., 82
   E. Barbieri et. al.: ibid., 82
   S. R. Ramee et. al.: ibid., 85
2. V. Russo et. a.: Proc. SPIE, 405 (1983) 21
3. H. Schmidt-Kloiber: German Patent Application DE 36 00 730 A 1 (1986)

Methods in Lithotripsy

H. Wurster

Richard Wolf GmbH, Pforzheimer Str. 24,
D-7134 Knittlingen, Fed. Rep. of Germany

1. Introduction

Urinary stones and Lithotripsy are two terms, which are mentioned together and have a long tradition in the medical field.

From the stone cutters in the middle age, who travelled around with their instruments to the first subrapubic approach using a chisel, the endoscopic mechanical lithotrites with optical viewing system, which are very painful to introduce, a newer method of lithotripsy using an electrohydraulic shockwave probe, which is inserted into the patient's body through the endoscope up to the most modern extracorporeal lithotripsy by shockwaves, was a long path of development (Fig. 1).

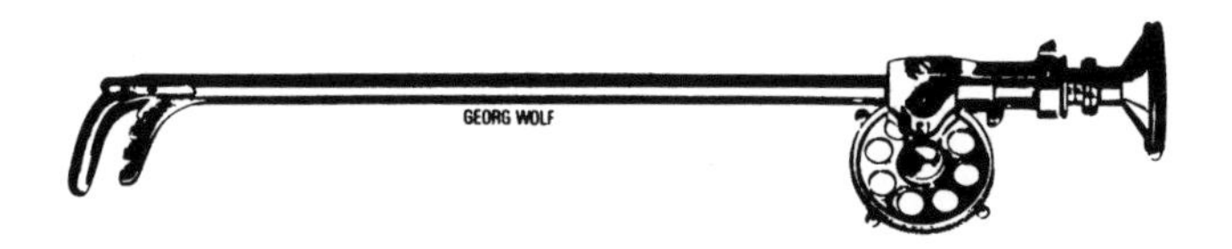

Fig. 1 Mechanical lithotrite with optical system

2. Electrohydraulic Lithotriptors

The electrohydraulic lithotripsy was invented by the Russians in 1965 and the technique was improved and refined to smaller and thinner probes handling less energy, so that they could be used also within the ureter.

The electrohydraulic lithotriptor is very effective, specially on bladder stones, where it is used endoscopicly with a 10 Fr. probe.

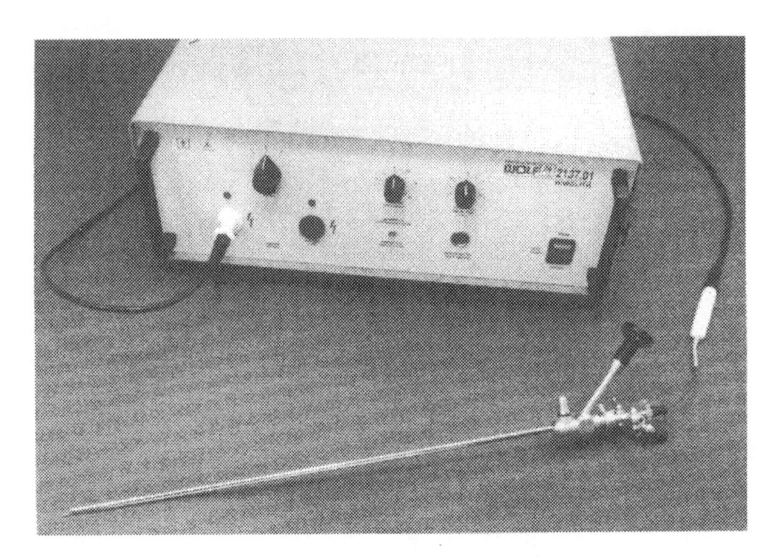

Fig. 2 Electrohydraulick shockwave generator with 3 Fr. probe inserted into an ureteroscope

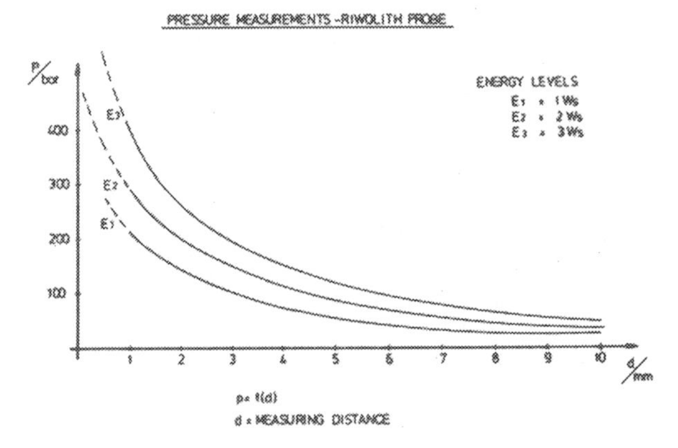

Fig. 3 Pressure versus distance at an electrohydraulic probe

The energy per shock can be increased up to 5 Joules and the repetition rate up to 40 shots per second (Fig. 2). The pulse width is approx. 1 μsec. For not destroying the hydrophones pressures could only be measured at a certain distance from the probes (Fig. 3). There are surely more than 1000 bars directly at the tip of the probe, which has to be in contact with the stone. This is very important for good stone disintegration. Some of the fragmented stones tumble around into the bladder and may do some irritation to the bladder wall with their sharp edges. The gasbubbles of the spark discharge are producing a second pressure pulse some milliseconds after the shockwave, which causes the movement. It takes much time to aim on each fragment and disintegrate it small enough for rinsing out. Therefore a combined method is used. Big stones are crashed with the electrohydraulic lithotriptor into several pieces and then an endoscopic punch forceps is used to disintegrate them so small, that they could be transported out by suction through the punch lithotrite's lumen.

Experiments have shown, that at the same energy level higher voltage is more effective than lower voltage due to the rise time. A pulse with a shorter rise time will give smaller concrements.

Using the electrohydraulic disintegrater in the ureter, the probes are as small as 3 Fr. and are used only at a low power setting and in single shot mode.

3. Ultrasonic Lithotriptor

Another method in Lithotripsy developed in 1970, is the ultrasonic lithotriptor, which was designed for bladder stones, but was not as effective as the electrohydraulic system for this purpose.

The come back for this ultrasonic unit was the percutaneous approach of kidney stones, where the electrohydraulic shockwave first seems to give harm to the soft kidney tissue.

The tip of the ultrasonic probe, which is oscillating longitudinal and transversal at a frequency of about 20 - 40 kcs is very harmless, if it touches the tissue. The ultrasound travels very easily through the tissue, which has nearly the same acoustic impedance as water and does therefore not damage the kidney tissue, if the power density is not too high. On the stone it acts like a drill and the hollow ultrasonic probes allow to suck off the stone dust and water to cool the probe additionaly. A continuous water irrigation system keeps the view clear for the endoscopist within the kidney pelvis. This unit is used worldwide for kidney stone treatment in connection with endoscopes (Fig. 4).

The ultrasonic intensity, which is applied is in the range of 3 W/qcm, which does not give damage to tissue, as animal experients showed.

The thinner probes of such a system down to 1 mm could be used within the uretero renoscopes to disintegrate ureteral stones as well (Fig. 5), (Fig. 6).

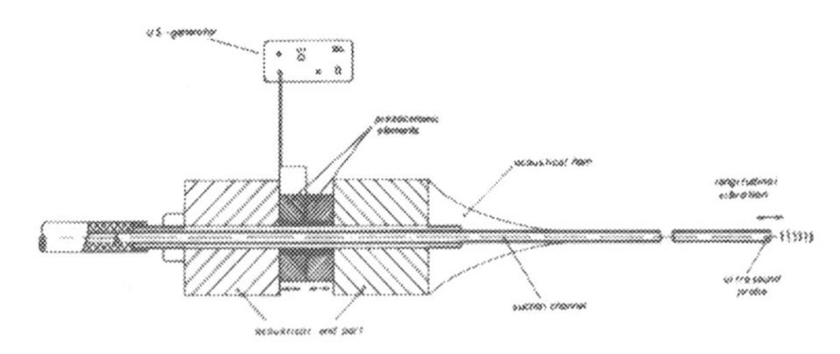

Fig. 4 Principle of an ultrasonic driven lithotrite, using a sandwich type transducer

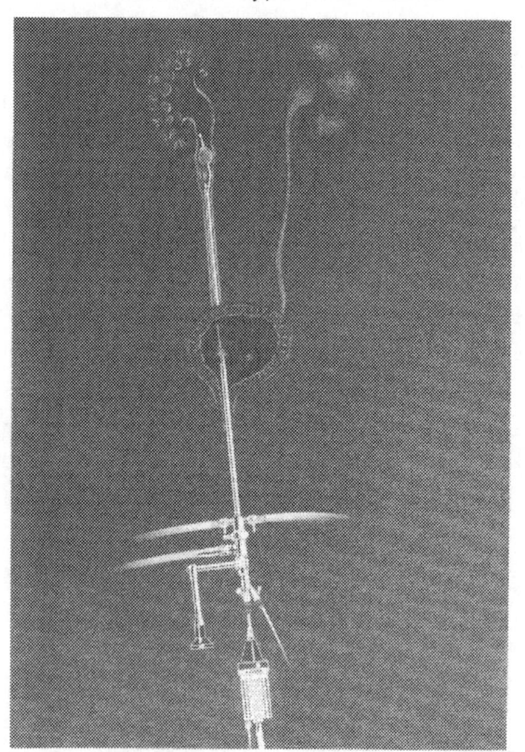

Fig. 5 Uretero renoscope with ultrasonic lithotriptor and blocking balloon

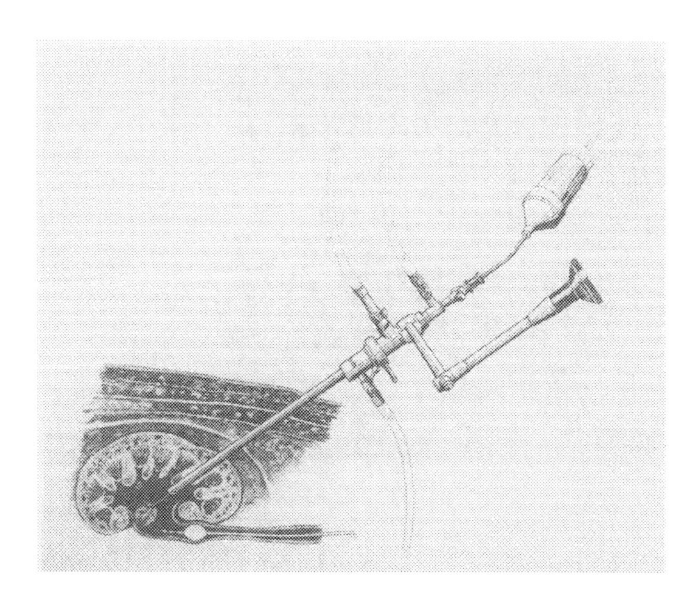

Fig. 6 Percutaneous approach by nephroscope

A dormia basket may be used to keep the stone in place, if it slides too easy within the ureter. Also a balloon by passing the stone and blown up in front of it to keep it in place that the ultrasonic probe can work on it. This system is very useful, specially in addition after extracorporeal shockwave treatment at "Steinstraße", when the stone concrements block the ureter.

4. Laser Lithotripsy

In the ureter also the laser lithotripsy with their thin fibers is a useful instrument. The flexible transmission system of the fibers with the outer diameter from 0,5 - 1 mm are very comfortable for the ureteroscopic procedures. Due to their small focal point and small fiber diameter, high energy density is achieved (Fig.7).

The small thickness and flexibility is the most advantage of the laser to fit into the smallest scope. The eroding effect due to small energy per pulse make the treatment safe. High repetition rate of laser pulses give reasonable destroying time for stone. There are no problems in treating all different types of stones. Harder stones need more time than softer ones. The ultrasonic probe is semirigid and power is decreased at the tip, if it is bent. Rigid application is recommended.

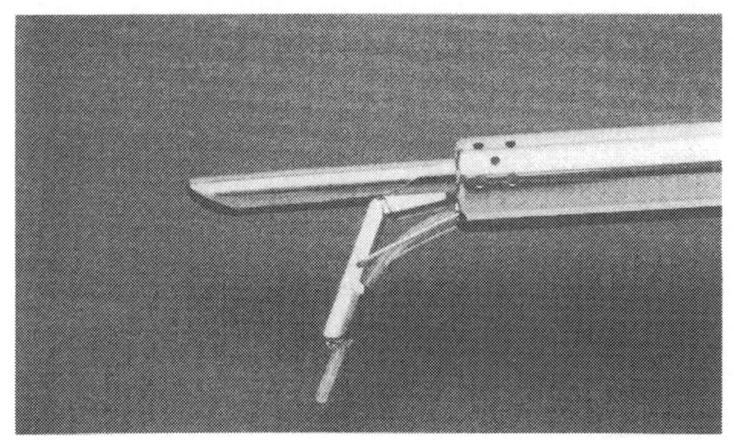

Fig. 7 Endoscope with deflecting system for laser fiber

5. Extracorporeal Lithotripsy

The latest and most inventory technique, which was developed in the last ten years, is the extracorporeal shockwave lithotriptor. Häusler in Saarbrücken first mentioned this effect in 1971 and showed cracking stones using shockwaves. Our development started in 1977 together with Häusler and Ziegler of the University of Homburg to design a bathtubfree shockwave lithotriptor, which could be applied from the outside to the patient having a spark gap array as shockwave generating element into an annular ellipsoid.

In 1980 a piezoceramic system was designed together with the University of Karlsruhe, Prof. Kurtze and Dr. Riedlinger. The principle of the system is, to guide the focussed shockwave into the patient's body from the outside and the focus is adjusted to the stone. The spark gap generated shockwaves into the bathtub is commercially used in clinics from 1983.

The difference between the spark gap generated shockwave and the piezo shockwave at this time is the aperture of the generator, which is larger at the piezo. The focal area therefore is smaller and the pressure density at the patient's skin is low, that the piezo unit works painfree and does not require anesthesia to the patient.

This is the key feature of the EPL (Extracorporeal Piezo Lithotriptor). The pressure pulses are made by discharging a high voltage on the piezoceramics. The rise time of the pressure pulses is in the range of 300 nsec. The pulse length will be about 1 microsec.

A typical pressure pulse is shown in Fig. 8.

The medium pressures are above 1000 bars into the focus. The focal area of the system is very small and the half pressure area is 3 x 9 mm.

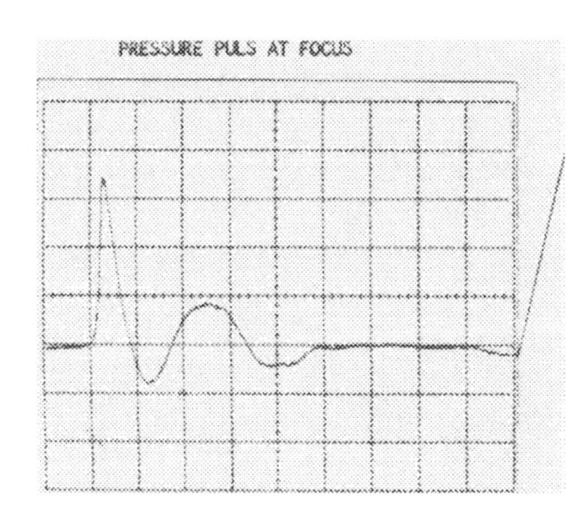

Fig. 8 Pressure pulse
1 V / 1 μsec. per div.

With the spark gap and ellipsoid the shockwave is created at the focus 1 of the ellipsoid and the reflected into the second focal point, which is inside the body (Fig. 9a), (Fig. 9b). At the piezo-system there is an ultrasonic pressure pulse generated and by propagation of the ultrasound towards the focal point the pressure increases and due to the nonlinearity of the water a shockwave formation is created before and after the focal area.

Fig. 9a Piezoceramic

Fig. 9b Spark gap

For locating of the stone an ultrasonic diagnostic scanner is used, which works as a sector B-Scan and shows the real time image of the stone. Such ultrasonic location has the advantage over x-ray, which is used in other lithotrites, that there is no radiation exposure to the patient (Fig. 10).

Based on those physical facts a unit was designed, named Piezolith 2300, which is a mobile unit, does not need much room and no special installation, an electric wall outlet, a watertap and a drain is sufficient to operate this unit.

The table plate could be shifted 70 cm footwards. To allow additional controlling of stone disintegration a x-ray-unit will be brought in position. In this position the table also could be used for other urological auxiliary procedures supplied with legholders and an urological drain.

The unit is used for destroying kidney stones, which are well located with the ultrasound and also for ureteral stones, whereare some problems in the middle part of the ureter, covered by bones, which will not allow visualization with ultrasound from the backside. In some cases it could be done in a prone position.

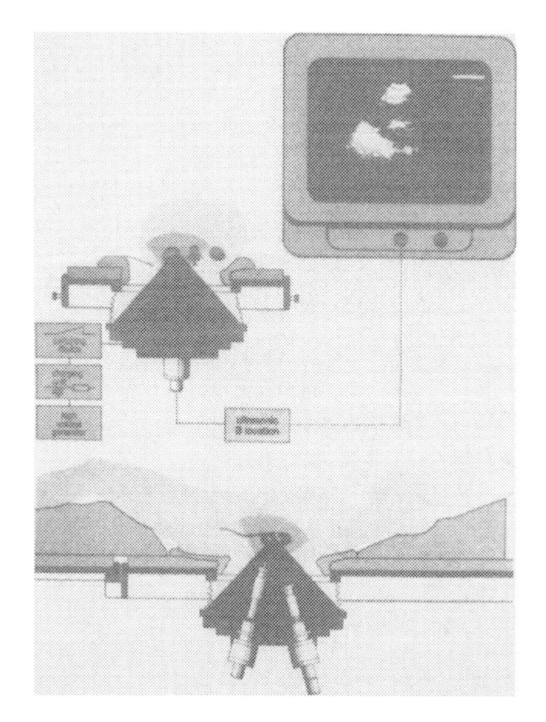

Fig 10 Principle of EPL-unit with ultrasonic location system

The location system has two scanners, which are tilted to allow good access to the kidney, specially for locating stones under the ribs and the intramural stones behind the symphysic bone.

The machine could be operated by one person, the urologist himself. There is no need of any other person, which makes the procedure very cost effective.

Treatment of stones could be done in most cases on an outpatient basis and also the multistage treatment is very effective to reduce the "Steinstraße". Big kidney stones or staghorn stones are treated in multiple sessions, where at the renal pelvis is started to allow passing of the concrements.

For a stone with 10 mm diameter 1000 - 1500 shocks are used.

Before the patient is placed on the table, with an hand held diagnostic scanner the stone is located to have good ultrasonic access to the stone, this point is marked on the skin. Now the patient is placed with this mark over the integrated ultrasonic scanning head and the image of kidney and stone is seen on the screen. The transducer is positioned under ultrasonic guidance, that the stone comes into co-incidence with the cross hair, which represents the focal point. The ultrasound runs continuously during the procedure.

The ultrasonic scanning head could be rotated at 90 ° and also lifted up and down to adapt the patient's anatomy. The shockwaves are released manually, pressing a buttom as single pulse or in the other mode continuously up to 2,5 Hz (Fig. 11).

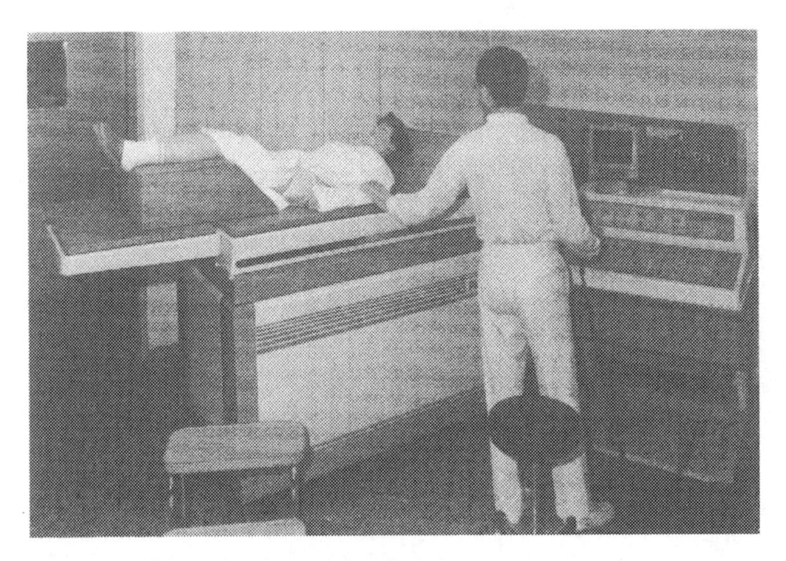

Fig 11 EPL-unit in operation

There have been no signs of arrythmia or extrasystoles with the Piezolith seen as it is noticed by the units having spark gap generators.

There are about 20 unit out in the field and worldwide more than 3000 patients have been treated. All of them without any anesthesia.

Only on small babies anesthesia is required to keep them in position.

This type of extracorporeal shockwave will also be used for gall stones in near future, there are no problems to disintegrate those stones.

6. Conclusion

Looking over the total methods of lithotripsy including the laser, it could be said, that each unit has its special features with its special application and advantages. In some cases different methods have to be used together and one supplements the other. So at staghorn stones the extracorporeal and the percutaneous method are used in conjunction.

By percutaneous method parts of the stone at the kidney pelvis are taken out, that there is room for the stone debris at following ESWL.

The "Steinstraße" is the field of lithotripsy, in which the laser has big advantage combined with the ureteroscope. Although it is competing with ultrasound and electrohydraulic, which are low cost units. Bile duct stones, which are fixed into the duct, could be treated only with laser or electrohydraulic. The thin laser fiber, due to their flexibility in conjuction with a flexible endoscope, is the method of choice. Disadvantage is that Q-switched and free running pulsed laser give different results in biliary and urinary stones.

The following table showes an overview about the different methods used today (Fig. 12).

Methods of Lithotripsy

method	urology			gastroenterology	
	bladder	kidney	ureter	gall bladder	bilary duct
blind mechanical instruments					
forceps	–	–	–	–	–
loops, baskets	–	–	⊝	–	–
endoscopic instrument (in connection with)					
forceps	+	⊝	⊝	⊝[1]	⊝
punchlitho	+	–	–	–	–
dormia basket	–	–	⊝	–	⊝
electrohydraulic	++	⊙	+	⊝[1]	⊝[1]
ultrasound	⊝	++	++	⊝[1]	–
dye laser / YAG-laser	–	⊝	++		+[1]
Extracoporal shockwave	–				
Spark gap / electromagnetic / piezo		++	+[illegible]	+[2]	+[3]

legend: – not used; ⊝ limited; + good; ++ very good

1) in experimental phase
2) only on cholesterine stones with chemical solution
3) with aux endoscopic procedure after treatment
4) not in the middle third of the ureter

Fig. 12 Comparison of the different methods in lithotripsy at the different organs

Devices for Intracorporal Laser-Induced Shock-Wave Lithotripsy (ILISL)

F. Wondrazek and F. Frank

MBB-Medizintechnik GmbH, Postfach 801168,
D-8000 München 80, Fed. Rep. of Germany

With high power lasers electrical field strengths of light above the threshold of optical breakdown can be generated. The amplitudes of the emitted shockwaves are high enough to fracture all kinds of stones deposited in the human body. Via a special device light pulses of 70 mJ and 20 ns generated by a q-switched Nd:YAG laser are coupled into a flexible optical fiber. The parameters of the laser system are optimized for the beam transmission in thin quartz-glass fibers of down to 0.6 mm in diameter. In the vicinity of the concrement the electromagnetic energy of the light pulse is converted into acoustic shockwave energy by means of special optical devices or opto-acoustic couplers. These systems are combined with miniaturized acoustic reflectors which direct the shockwaves towards the calculus. An essential feature of these devices is their endoscopic applicability.

1. Introduction

Among the various noninvasive methods for the fragmentation of stones, laser lithotripsy is becoming increasingly important. A variety of laser systems, e.g. dye-, excimer- and Nd:YAG laser, are being investigated to determine their action upon stone material. Markedly different effects occur, for example thermal or mechanical fragmentation, depending on the different wavelengths and light intensities of the laser systems as well as the varying chemical composition of the concrements.

To be independent of the photochemical properties of the concrements, we have developed a system which converts the electromagnetic energy of a laser light pulse intracorporally into the acoustic energy of a shockwave. The lithotriptor is based on a specially developed, q-switched Nd:YAG laser whose high power light pulses are coupled into a thin flexible quartz fiber by means of a novel device. In the vicinity of the targeted concrement the light transmitted via the fiber generates a plasma state in a liquid by means of a focusing device or an opto-acoustic coupler. A miniaturized acoustic reflector directs the resulting shockwave on to the stone.

2. Laser System

The q-switched Nd:YAG laser system is designed to deliver light pulses of approximately 20 ns in duration. Spikes in the temporal variation in intensity measured with a silicon diode of high time resolution indicate the presence of several longitudinal modes. The pulse energy is continuously variable up to 100 mJ. The repetition frequency can be set from single-shot operation to 100 Hz.

Measurements of the spatial intensity distribution of a light pulse with a silicon diode-array show the presence of several transversal modes (Fig. 1). The multimode-operation leads to an approximately rectangular beam profile. As a result of mode interference, however, considerable intensity spikes may develop. The rectangular intensity distribution guarantees an uniform strain on the cross-section of the quartz fiber. TEM_{00}-mode operation with its gaussian beam profile stresses the central part of the fiber much more than the periphery. Intensity spikes in the cross-sectional profile or in the temporal behaviour limit the maximum transmittable pulse energy considerably. By increasing the pulse energy and with it the mean power density level, the intensity spikes rapidly exceed the damage threshold of the fiber material.

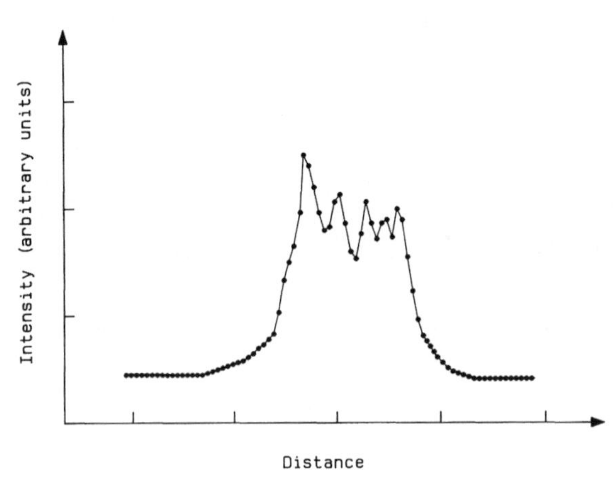

Fig.1. Typical spatial intensity distribution of a laser light pulse

3. Fiber Coupling

Quartz fibers with core diameters of 600 µm or 1000 µm and external diameters of 950 µm and 1400 µm respectively are used. With the Nd:YAG laser at 1.064 µm the attenuation is insignificant.

The surface quality has a decisive effect on the damage threshold of the light guide. The irradiation of dirt particles, scratches or other damage with intense light pulses may induce irreversible changes which finally result in the destruction of the surface. Better results can be achieved by cutting the fiber than by polishing the fiber surface.

It is also very important that the coupling of the laser light is effected parallel to the axis of the light guide. If there is an angle between the incident light beam and the fiber axis, the intensity is increased in the vicinity of the point of impact by reflection at the convex surface. As a consequence a multiphoton-ionisation and avalanche-breakdown in the fiber material may occur.

Inhomogenities and impurities inside the fiber material may act as nuclei for a light induced avalanche-breakdown due to a local increase of the linear or nonlinear absorption.

At sufficiently high light intensities the so-called self-focusing effect as a damaging mechanism has to be taken into account. As is well known, the index of refraction depends on the light intensity. Consequently a nonuniform intensity distribution across the fiber section leads to a varying power of refraction and thus to a lens effect of the homogeneous quartz fiber. The self-focusing effect can be avoided by an uniform distribution of the light intensity across the fiber section.

High light intensities give also rise to nonlinear effects such as stimulated Raman scattering and stimulated Brillouin scattering. The wavelength of the incident light is shifted. Due to the wavelength dependent absorption such shiftings may cause irreversible damage in the fiber material.

All these various damaging mechanisms can be avoided by using a novel fiber coupling device for high light intensities (Fig. 2). By means of an optical system the light pulse is focused into a breakdown-medium. The focal spot-diameter and the breakdown-medium are selected in a way that intensity spikes exceeding the damage threshold of the fiber induce an optical breakdown before entering the fiber. Pulses with spikes of such intensity are in this way absorbed by the high pulse filter. The light guide is located in the divergent beam behind the focal waist. The diameters of fiber and light beam are matched. A too narrow beam diameter results in unnecessarily high light intensities at the fiber surface. A beam diameter exceeding the fiber diameter causes diffraction with intensity peaks at

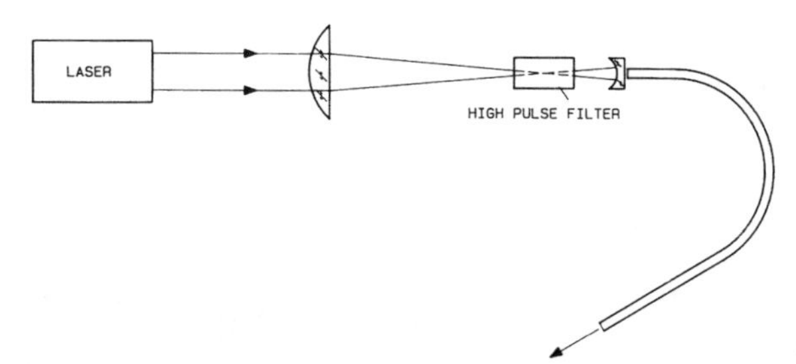

Fig.2. Fiber coupling device for high light intensities

the fiber edge. In addition, the coating and the fiber plug may be damaged by the intense light. By means of a negative lens in front of the light guide the divergence of the light beam inside the fiber is increased. On the one hand the self-focusing effect is thus prevented, on the other hand a rapid homogenisation of the intensity profile occurs because of frequent reflections at the fiber surface.

With this device homogeneous light pulses with energies of up to 70 mJ, at repetition rates of up to 30 Hz are received at the exit of a 600 μm fiber. This corresponds to an average power of more than 2 watts. The peak power density in the fiber exceeds 1 GW/cm². Using a 1000 μm fiber the transmittable pulse energies are significantly higher.

4. Generation of Shockwaves

For the generation of the shockwaves at the distal end of a fiber various methods have been investigated. All these systems are designed to be integrable in standard rigid or flexible endoscopes.

The divergently emitted light from the fiber can be concentrated by means of a focusing device into a fluid surrounding the concrement (Fig. 3). The light intensity in the focal region exceeds the breakdown threshold of the fluid, thus producing an optical breakdown. An elliptical acoustic reflector concentrates the shockwave energy released in one focal point of the ellipse to the second focal point which is located in the direct vicinity of the targeted concrement. The developed focusing system causes an aberration-free reduction of 10:1 of the fiber's outlet aperture. The lens next to the optical breakdown is made of a special material to withstand the high compressive stress. Its surface acts as a part of the acoustic reflector. The maximum external diameter in combination with a 1000 μm fiber is 4.5 mm. The focusing system for the imaging of a 600 μm fiber has, together with the flushing channel, an approximately elliptical cross-section with the overall dimensions of 2.8 x 3.5 mm. To reach the threshold for the optical

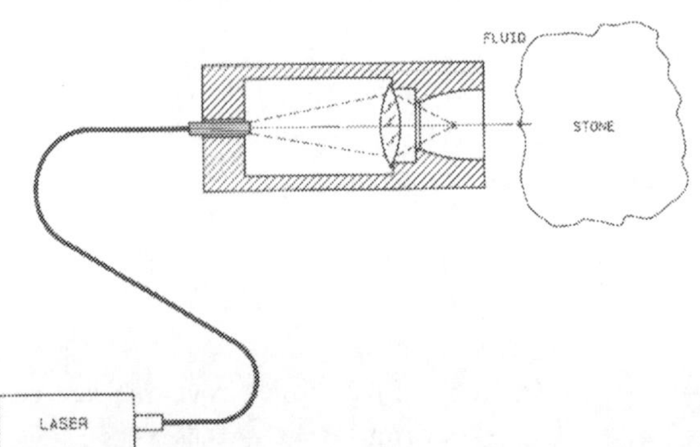

Fig.3. Shockwave generation by the focusing technique

breakdown in distilled water, a pulse energy of approximately 10 mJ corresponding to an intensity of about 8 GW/cm^2 is necessary. An uric acid stone of approximately 1 cm^3 in volume could be destructed easily with the focusing device (Fig. 4). After about 350 pulses with an energy of approximately 45 mJ per pulse the stone is disintegrated in dust-like particles and pieces of 2-3 mm in dimension.

Fig.4. Uric acid stone after laser induced shockwave lithotripsy

One can get a very simple focusing device by spherically grinding or melting the distal end of a quartz fiber. Due to lens abberations as well as the very small object distance the focusing effect is not very good. The threshold energy for the optical breakdown in distilled water with a 600 µm fiber is approximately 40 mJ.

Another technique we use to produce shockwaves avoids focusing of light by bringing an absorbent material into the divergent laser beam. As a result of the high linear or nonlinear absorption the breakdown threshold is decreased and a plasma state followed by a shockwave is formed at the surface of the absorber. In consideration of the durability a hard metal wire is used as absorber. The zone between light guide and absorber acts as acoustic reflector. Light guide and absorber are surrounded by a fluid. The system has a maximum external diameter of 1.9 mm. The opto-acoustic coupler is placed directly on the concrement. The geometry of the system and the absorber material determine the threshold.

5. Conclusion

Each of the distinct distal application systems for the generation of shockwaves - the focusing device, the ground quartz fiber and the opto-acoustic coupler (Fig. 5) - has advantages and disadvantages regarding its dimension, threshold, lifetime and

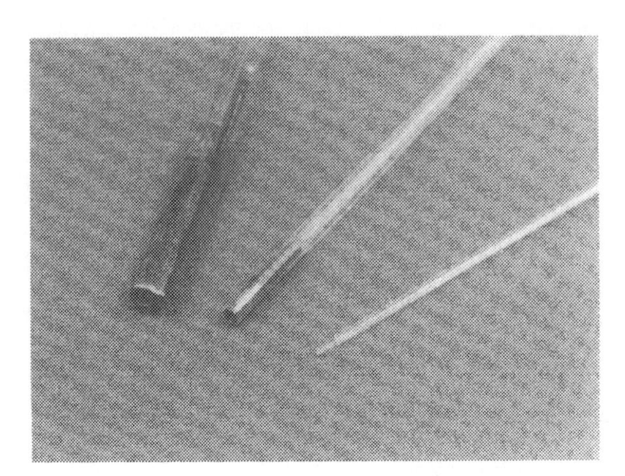

Fig.5. Distinct distal application systems for shockwave generation

handling. As regards the application it becomes apparent that the focusing system is best suited for the percutaneous litholapaxy, the opto-acoustic coupler for the destruction of ureter stones and the bare fiber for the treatment of gallstones.

A More Effective Method of in vitro Urinary Stone Disintigration Using the Nd:YAG Laser

R. Friedrichs[1], *R. Poprawe*[2], *W. Schulze*[2], *M. Wehner*[2], *W. Schäfer*[1], *and H. Rübben*[1]

[1]Department of Urology RWTH Aachen, Pauwelsstr., D-5100 Aachen, Fed. Rep. of Germany
[2]Fraunhofer-Institute for Laser-Engineering and -Technology, Steinbachstr. 15, D-5100 Aachen, Fed. Rep. of Germany

A tube system for urinary stone disintigration which contains as essential parts a water chanal and a Nd:YAG laser beam is presented. The calculi are sucked to this system by a pump, thus being in direct contact with the tube in 90% of the time. The entrance of the tube system has two functions: a) the laser beam leaves the tube via this entrance and hits the calculus and b) the fragments from the calculus splint off and are immediately sucked to the water chanal via the entrance of the tube system. Thus the stone is completely disintigrated step by step. Only fragments equal in size with the entrance or smaller are sucked to the water chanal. By variation of the entrance size of the tube system, the fragment size of the urinary calculi can be determined before starting disintigration. The method described by us allows stone disintigration into very small uniform particles (100 - 200 µm, intensity of laser radiation I= 10^9 W/cm^2).

Introduction

The first application of laser radiation for urinary stone disintigration has already been described in the 1960s (1). Meanwhile a variety of different laser systems have been described in vitro and have already been introduced into clinical practice (2-8). Up to now there is however no ideal laser system which fulfills all requirements for ideal endoscopic treatment set up by Miller (9): safety to the patient and operator, attainment of complete calculus destruction and removal, the possibility of a reload or multiple shot facility, rapid action with minimal bleeding, small diameter probe, preferably flexible, variable power settings, cost effectiveness and reliability of equipment.

The purpose of this paper is to present an in vitro tube system which allows determination of fragment size before starting urinary stone disintigration.

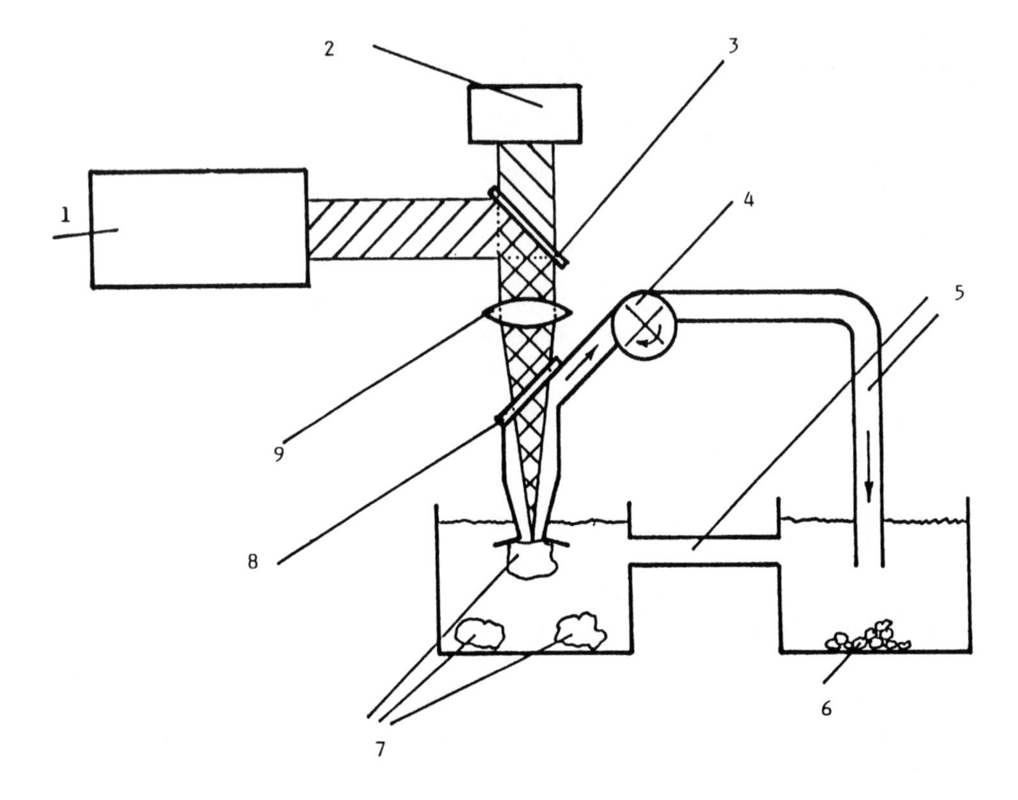

Fig. 1: Experimental set-up for disintigration of urinary calculi (schematically)

1) Laser
2) Optical process control
3) Beam coupling
4) Pump
5) Circulation
6) Fragments
7) Urinary calculi
8) Window
9) Focussing optics

Patent application number P 37 11 086.1

Method and Material

Figure 1 shows the experimental set up for disintigration of urinary calculi. Laser radiation emerging from a source (Neodym YAG laser, dye laser, excimer laser respectively) (1) is brought to the experimental set-up. Via a reflecting mirror (3) coaxial coupling of the laser radiation to the optical beam pass for process control (2) is achieved. In doing so, simultaneous optical control of the disintigration process and processing the calculi is possible. The calculi (7) are sucked to the opening of the device by a pump (4). Simultaneously fragments of the calculi are transported through a circulation system (5), such that the fragments (6) are deposited in a corresponding container. The liquid in the circulation system - usually physiologic NaCl solution - is separated from the optical beam pass by a transparent window (8). The laser radiation is focussed by a microscope lens (9) through the window into the liquid of the circulation system. By variation of the opening of the device (7, 4) the fragment size can be determined before starting disintigration.
Urinary calculi of different chemical composition and size are treated with the experimental set-up described above.
Laser radiation is emerging from a Q-switched Nd:YAG laser (pulse duration $\tau = 15$ ns, intensity $I = 10^9$ W/cm^2, pulse energy $E = 25$ mJ, focal radius $2r_F = 500$ µm and laser frequency $\nu = 20$ Hz).

Results

Urinary calculi of different chemical composition and of different size can be successfully disintigrated by the described experimental set-up. By variation of the entrance size of the tube system the fragment size can be determined before starting disintigration (minimum fragment size 100 - 200 µm). The urinary calculi are in direct contact with the tube system in about 90 % of the time.
The application of the experimental set-up results in a uniform fragment size. Figure 2 shows the result after treatment of a urinary calculus (calcium oxalate) with a Q-switched Nd:YAG laser (parameters mentioned above). Fragments of different size occur. Figure 3 shows the result after disintigration of a urinary calculus (calcium oxalate) in the described experimental set-up under the same parameters of laser radiation as in figure 2. Now fragments of a nearly uniform size occur. - Successful urinary stone disintigration in the described experimental set-up is also possible with laser radiation emerging from a dye laser or excimer laser .

Fig. 2: Fragments of different size after Nd:YAG laser treatment (intensity I= 10^9 W/cm^2)

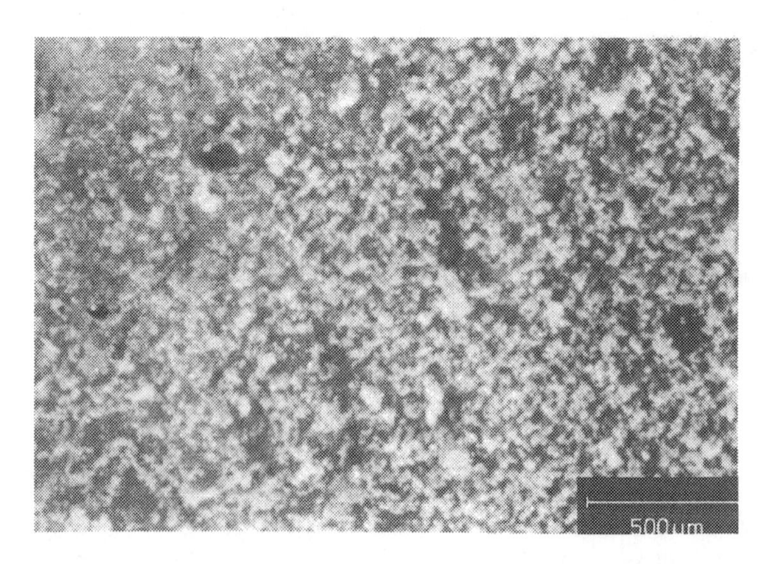

Fig. 3: Fragments of uniform size after Nd:YAG laser treatment in the described experimental set-up (intensity I= 10^9 W/cm^2)

Discussion

To our knowledge this is the first description of a method which allows determination of fragment size before starting disintigration. Furthermore the method incorporates three major demands important for successful application:

1. positioning of the urinary calculus
2. automatic repositioning after laser treatment
3. coaxial set-up of laser radiation and optical process control (10-12).

The further development and the introduction of the described experimental set-up into clinical practice may have several advantages: a faster and more effective urinary stone disintigration, no residual stone chips left and the development of endoscopic instruments which are smaller in diameter.

Up to now a standardized terminology in the new field of laser lithotripsy has not yet been introduced. This fact makes the comparison of results and the discussions between different groups difficult. It is concluded that publications about laser lithotripsy should give all important characteristics of the laser radiation used including intensity (I , W/ cm^2) and focal radius ($2r_F$, µm).

One should not forget that laser treatment always requires invasive therapy (percutaneous puncture and treatment through the nephrostomy tract or cystoscopy and ureteroscopy). The alternative to invasive treatment is the the non-contact ESWL which is available today at more than 2o centres covering our country. More than 170,000 patients have been successfully treated up to now worldwide. Any new development in the field of laser lithotripsy will have to prove its superiority over the non-contact ESWL which can also be used for urinary stone disintigration in the ureter. Although the close cooperation between physicians and engineers offers new ways of treatment, one should always ask critically which method is really the best for a patient.

Literature

1. W.P. Mulvaney, C.W. Beck: J. Urol. 99, 112 (1968)
2. J. Pensel, F. Frank, A. Hofstetter, K.H. Rothenburger: Proc. 4th Congr. Las. Surg., Tokio (1981)
3. H. Schmidt-Kloiber, E. Reichel, H. Schöffmann: Biomed. Technik 30, 173 (1985)
4. R. Hofmann, W. Schütz: Urol. (A) 23, 181 (1984)
5. A. Hofstetter, F. Frank, E. Keiditsch, F. Wondrazek: Laser 1, 155 (1985)

6. G. Watson, S. Murray, S.P. Dretler, J.A. Parrish: J. Urol. 138, 195 (1987)
7. G. Watson, S. Murray, S.P. Dretler, J.A. Parrish: J. Urol. 138, 199 (1987)
8. S.P. Dretler, G. Watson, J.A. Parrish, S. Murray: J. Urol. 137, 386 (1987)
9. R.A. Miller: World J. Urol. 3, 36 (1985)
10. R. Poprawe, E. Beyer, G. Herziger: Proc. GCL 1984, ed. by A.S. Kaye, A.C. Walker (Adam Hilger Ltd. Bristol, 1985) p. 67
11. R. Friedrichs, R. Poprawe, R. Kohnemann, W. Schäfer, H. Rübben: Laser/Optoelectronics in Medicine, ed. by H. Waidelich (Springer, Berlin, 1988) in press
12. R. Friedrichs, R. Poprawe, W. Schäfer, H. Rübben: Verh. Dt. Ges. Urol., 39, in press (1988)

Shock Wave Excitation by Laser Induced Breakdown (LIB): Pressure Measurements with a PVDF Needle Probe

M. Köster, S. Thomas, J. Pensel, M. Steinmetz, and W. Meyer

Medizinisches Laserzentrum Lübeck GmbH,
Peter-Monnik-Weg 9, D-2400 Lübeck 1, Fed. Rep. of Germany

To achieve shock wave excitation by laser induced breakdown (LIB) by means of a Q-switched Nd:YAG laser (s=1064 nm, Spectron/MBB) different couplers and fiber systems can be used, THOMAS /1/. The fast dynamic changes of pressure can fragment urinary or biliary stones into tiny particles. The values known for a minimal static pressure needed for stone destruction vary from 80 bars for magnesium-ammoniumphosphate stones to 22 bar for gall-stones consisting of cholesterin and bilirubin, HÄUSLER /9/. The transformation of light energy by dielectrical breakdown into acoustic energy has a ratio of about 6 % (VOGEL /2/) and can be explained as follows: Similar to the electrical breakdown in the air during thunderstorms shock waves in form of pressure waves are produced by LIB in fluids. The laser light provokes formation of a plasma bubble in form of hot gas which consists of free electrons and ions, MEYER /3/. The tendency of the hot gas to expand on the one hand, and the hydrostatic pressure of the fluid on the other hand, leads to an oscillation of the bubble.

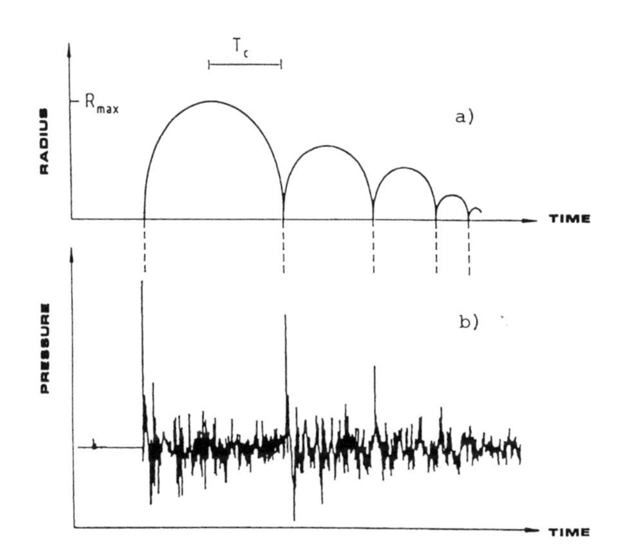

Fig. 1: radius-time-curve (a) and pressure-time-signal of oscillating bubble in water (b) T_C time of collaps (VOGEL /2/)

Each expansion of the hot plasma with ultrasonic speed produces a shock wave resulting in an energy loss of the bubble. Bubble dynamics and pressure distributios are completely altered by a solid surface.

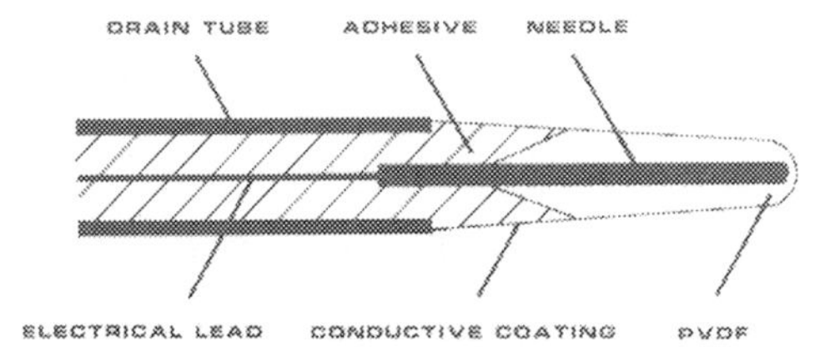

Fig. 2: construction of a PVDF needle hydrophone (schematic) MÜLLER /6/

Pressure measurements of the shock waves require a hydrophone with fast pressure change resolution and high mechanical stability which also does not disturb the pressure field. The polyvinyl-difluoridin-needle probe used in our experiments (IMOTEC /7/) guarantees high spatial resolution below 500 µm, because only the tip of the needle is the sensitive area. A small portion of the PVDF layer is piezo-electric and has a linear behaviour up to 300 bar, MÜLLER /6/. The rise time of our signals is about 150-200 ns and therefore slow compared with pressure changes within about 10 ns (VOGEL /2/) or even 4 ns (REICHEL /5/). Therefore the signal represents only one part of the real pressure. Additionally, the calibration of the probe could not be revised exactly. In consequence the following values of the pressures are less than the true values and represent a quantitative description. Pressure and pressure profile were measured with different couplers and systems fixed in a holder regulated by x-y-z-

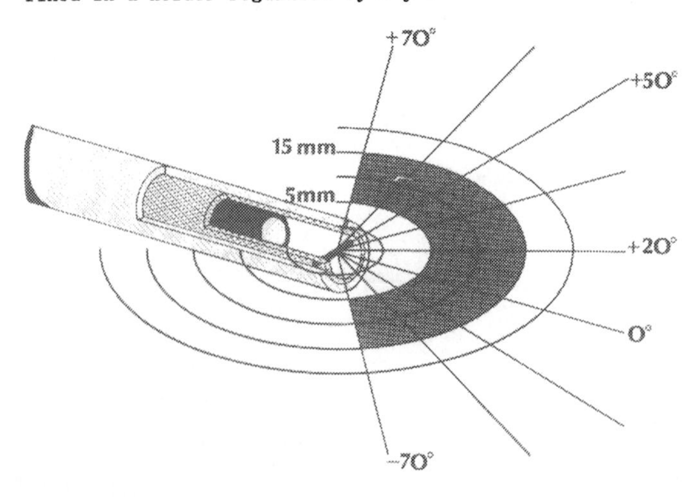

Fig. 3: plane of measurements with optomechanical coupler (THOMAS /1/)

micromanipulators. A 600 µm PCS fiber with an optomechanical coupler on the tip was brought in front of the pressure probe, so that pressure measurements could be done at different angles (-70° to +70°) and distances (3-20 mm) in one plane. To protect the needle probe from erosion by cavitation, the distance between hydrophone and the tip of the coupler was limited to 5-15 mm. The surrounding fluid was physiological salt solution to simulate clinical situation. This experimental set-up is mainly designed to measure pressure profiles of couplers with no need of stone surface for dielectrical breakdown. The induced plasma bubble grows between fiber tip and the metallic bar of the optomechanical coupler. The parameters in the pressure measurements are the distance fiber/bar and of course the position of the bar in relation to the plane (vertical/horizontal). The signals of the needle probe were recorded by a digital 100 MHz storage oscilloscope. Fig.4 shows a pressure signal of the optomechanical coupler. About 90 µs after the laser's trigger pulse the pressure signal of the shock waves after the breakdown appears. The pressure signal of the collaps follows 250 µs later. Because of spatial restriction inside the coupler an oscillation of the plasma bubble is suppressed and therefore no further pressure signals of shock waves are visible. For the following investigations the pressure maxima were used. With a calibration of 0.7mV/bar (IMOTEC /7/) you can calculate a pressure of about 14 bars and considering the time limitation of the signal the real pressure can be estimated - using signal's width and rise time (VOGEL /2/) - as exceeding 30 bars.

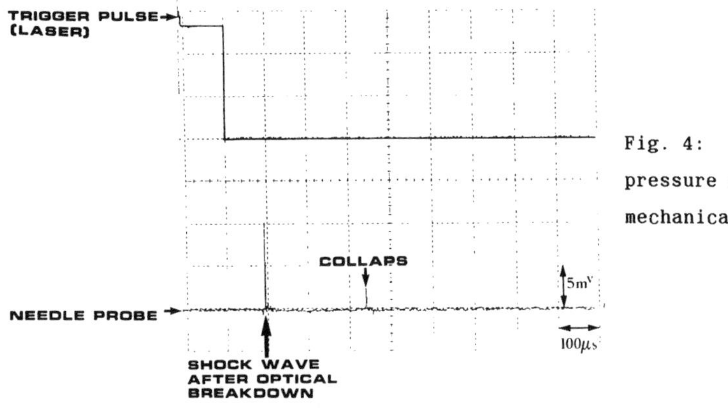

Fig. 4:
pressure signal of opto-mechanical coupler

The pressure values as a whole form a profile of angle and distance depency. As mentioned before, the angle varies from -70° to 70° and the distances from 5-15 mm. The maxima pressure are found in forward angles. Assuming the pressure to descend as described in literature in indirect proportion with the radius - like a hyperbola - one can estimate the pressure in front of the coupler's tip to be 300-400 bar. The low pressure between +/-10° - 20° can be explained with the position of the bar. It casts a shadow - or precisely - deflects the shock waves. A cut through this graph helps to visualize the spatial pressure characteristics.

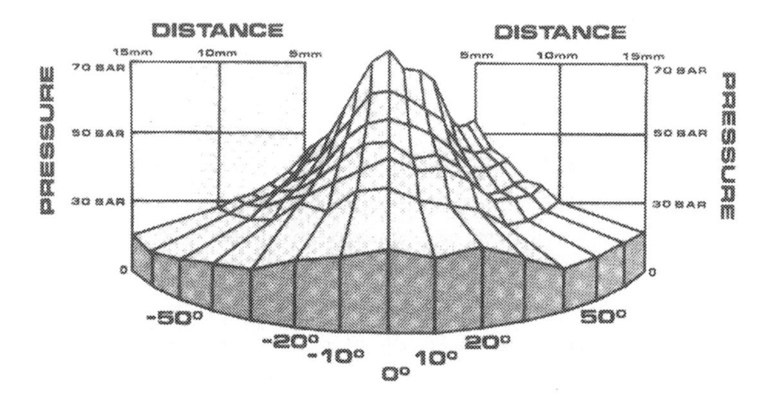

Fig. 5: pressure profile of optomechanical coupler

Measurements with a horizontal bar show a minimum of pressure around 0 degree. No spherical-symmetrical pressure distribution is revealed.

Fig. 6:
different angles
versus plasma at
5 mm distance (bar
vertical to plane)

Further pressure measurements investigated the influence of different distances between the fiber tip and the bar and during 6000 or 7000 shots. The distances varied from 3-4 mm while the laser energy (at the fiber's end) was fixed (at 27 mJ). The differences in pressure measured at a fixed location are due to the increasing volume which is useful for a bigger plasma bubble with higher pressure, LAUTERBORN /8/. The decrease of pressure during usage can be explained as follows: Due to the shock waves' pressure and thermal stress the surface of the fiber is damaged and an increasing amount of laser light is scattered and lost for the dielectrical breakdown. The initial pressure diminishes to about 40 - 50 % during 6000 shots. A detailed information about the optomechanical coupler and its durability is given in THOMAS /1/.

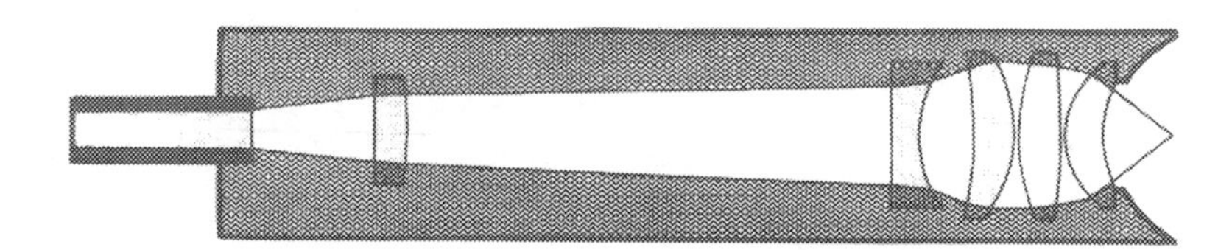

Fig. 7. optical focussed system

The next coupler consists of a system of lenses to focus the light delivered from the fiber end. A little fluctuation of the distance between lens and place of breakdown occurs due to smallest pollutions as seed for the breakdown and a cloud of microplasma (CHAHINE /4/) can be generated. The free bubble`s oscillation with a spherical distribution might be transformed into a cylindrical symmetry, REICHEL /5/. The pressure measurements were restricted by the coupler's short life time of about 100 shots (lenses were metallized and the focus disappears). They showed a pressure of 40-50 bar at 5 mm distance independent of the investigated angle within the limits of error (energy at fiber tip was 40 mJ). Two other systems to fracture stones are a spherical polished 600 µm HCP-fiber and a bare 200 µm AS-quarz fiber used with a flash lamp pumped dye laser, MEYER /3/. Especially the dye laser system cannot reliably produce a free breakdown in fluid with same energies as the couplers mentioned before. Like the optomechanical coupler additional electrons delivered by solid material are essential. The breakdown occurs on the stone's surface and only an unknown part of the shock waves is radiated into the salt solution. For comparable data the experimental set-up had to be changed. The fiber and probe were placed at known angles and distances to a surface of a stone model made of gypsum and different metal sheets. However, the surface of the model changes very soon after a few shots because of erosion the bubble disappears in cavities. The measured pressure in maximum of 60 bar at 5 mm distance using 25 mJ laser pulses and a metal determine only the trend. Measurements with small stones or metal sheets even show a vibration of the whole stone after a shot. To have a visual idea of the effects of a surface near a bubble fig. 8 from literature is representative.

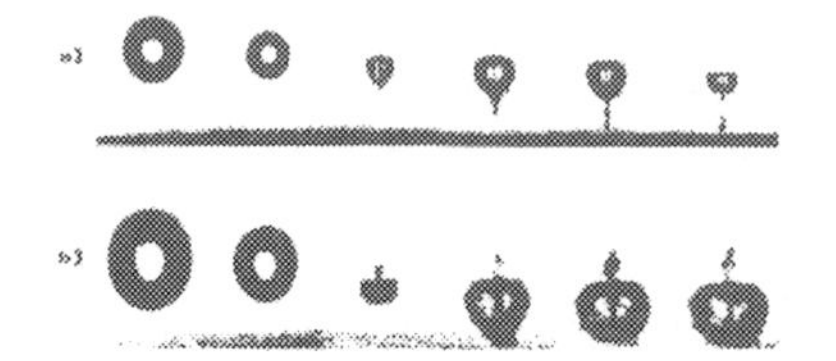

Fig. 8:
(VOGEL /2/)

The pressure waves radiated into the fluid are not spherical due to changed forms of collaps. Water-jets directed to the surface or even counter-jets with high velocities can occur. Although the shock waves' pressures of the different systems are sufficient to fracture urinary or biliary stones into tiny particles, they are probably only a part of the mechanism of destruction and further investigations have to be done to optimize the different systems.

1. Thomas S.: Optomechanical Coupler for LISL, in this book
2. Vogel A.: Opt. u. akust. Untersuchungen der Dynamik lasererzeugter Kavitationsblasen naher fester Grenzflächen, Dissertation Göttingen, May (1987)
3. Meyer W.: LIBS of Kidney Stones, in this book
4. Chahine G.L.: Chem.Eng.Commun. Vol.28 (1984) p.355
5. Reichel E. et al.: Laser 3 (1987) p.177
6. Müller M.: Acustica 58 (1985) p.215
7. data sheet of PVDF Hydrophone: IMOTEC GmbH, Paulinenstraße 136, D-5102 Würselen
8. Lauterborn W.: Acustica 20 (1974) p.51
9. Häusler E.: Amphora 4 (1975)

Shock Wave Detection by Use of Hydrophones

H. Schöffmann, H. Schmidt-Kloiber, and E. Reichel

Institut für Experimentalphysik, Karl-Franzens-Universität Graz,
Universitätsplatz 5, A-8010 Graz, Austria

1. Introduction

Piezoelectric polyvinylidene-fluoride (PVDF-) foils are nowadays commonly utilised as transducer materials in shock wave detectors. The only few um thick foils are excited in the thickness vibrational mode where the resolution in time is sufficiently high for ns phenomena. PVDF is only partially crystalline and therefore flexible as well as acoustically (nearly) transparent. Its piezoelectric constant amounts to 15 pC/N, yielding a sensitivity high enough for shock wave measurements without additional amplifiers.

Nontheless the time-resolved investigation of underwater shock wave sets some severe problems: The duration of laser induced shock waves is in the order of some 100 ns with details well below that, which is comparable to the signal travel time along the cable. In that case electrical reflections due to impedance mismatch between cable and input impedance of the measuring device are no more negligible, as they strongly change the shape of the pulse. /1/.

Statistically varying locations of the laser induced breakdown within the focal region imply temporal variations. Therefore no sampling devices are to be used for shock wave measurements, but ideally the fastest storage oscilloscopes, which are nowadays available.

Exact direct and non-static calibration methods do not exist for shock wave hydrophones. All applied methods need a lot of calculations and start from assumptions which are experimentally hard to prove. When a hydrophone is exposed to a shock wave, the pressure gradients are so high, that the behaviour of PVDF or any other transducer material seems to be unpredictable. So one has to apply different calibration methods and in comparison we might find the order of the shock wave pressure.

2. The Electrical Signal

One and the same shock wave will produce completely different electrical signals, depending on the acoustical and electrical parameters of the hydrophone and of the input impedance of the oscilloscope. /1/.

The shock wave compresses the piezoelectric foil, thereby changing its net dipole moment. Displaced charges appear on the electrodes, which are proportional to the instantaneous average stress inside the foil. During compression a part of the charge recombines over the input impedance. When the shock wave has passed, the foil takes its original dimensions thereby inducing a current of opposite sign. The smaller the impedance, the shorter the recombination time for a given quantity of charge. Therefore only with very high impedance, where no recombination of charge occurs at all, the time history of the voltage is proportional to the shock wave pressure.

On the other hand at ideal low impedance conditions - the recombination time is short compared to the pulse duration - the voltage is proportional to the temporal derivative of pressure.

Whenever both conditions are partly satisfied, the time history of the voltage can hardly be interpreted in terms of pressure. Typically a positive peak is followed by a negative voltage, which then, however, does not represent a negative pressure.

3. Further Parameters Influencing the Output Voltage

The thickness of the transducer determines the rise time of the electrical signal. A shock wave produces displaced charges as long as it penetrates the PVDF-foil. The voltage will rise till the shock wave reaches the backside of the transducer. Hence, the shortest measurable rise time is given by the transition time t, which can be calculated from the thickness of the transducer d and the shock wave velocity c within the transducer. A rise time of 4 ns results from a 9 um PVDF as c is about 2000 m/s.

A small active area is necessary for a high spatial resolution, as well as to avoid pulse broadening by differently long paths of the shock waves. Furthermore possible radial vibrations of the extended transducer are not picked up by the small active area.

As the PVDF-foil is surrounded by water, displaced charges may recombine due to conductivity of the fluid. The amplitude of the electrical signal is thereby reduced by approximately 20 per cent using tap water instead of distilled. At the worst no high impedance conditions are realizable, which can lead to misinterpretations of the temporal shock wave profile.

4. Experimental Results

All experiments were carried out using a Q-switched Nd:YAG laser, with a pulse duration of 20 ns at a wavelength of 1064 nm. The beam was focussed into distilled water by means of a simple biconvex lens. In the focal region microplasmas are formed, their number depending on the laser energy, on the focusing system, as well as on the distribution of contaminations. The time resolution of the hydrophone is sufficient to determine the plasma dimensions and the number of shock wave generated by one laserpulse.

4.1 Plasma Dimensions

At low energies (some mJ at the laser induced breakdown) a single plasma is produced, its location, however, fluctuates within 500 um. If we raise the laser energy up to some 10 mJ, we find a chain of microplasmas. At still higher energies (over 200 mJ at the LIB) the whole focusing cone is filled with tiny plasmas.

The experimental results were mainly obtained when the plasma forms a chain with a length of some mm and a breadth of 50 um. Because of this non-spherical geometry not only different shock wave forms will appear in laser beam direction and perpendicular to the laser beam, but also different peak pressures.

4.2. Shock Wave Shapes

Every microplasma is starting point of a spherical shock wave with different energies. These shock waves superimpose constructively in the direction perpendicular to the laser beam, forming a sharp shock front. (See fig.1). In laser beam direction, the shock waves are separated because of different starting points, thereby forming a "step-like" rise with lower peak pressures, but longer overall duration. As the number of plasmas as well as their locations vary statistically, the shock wave shape differs from laser pulse to laser pulse, as seen in fig.2.

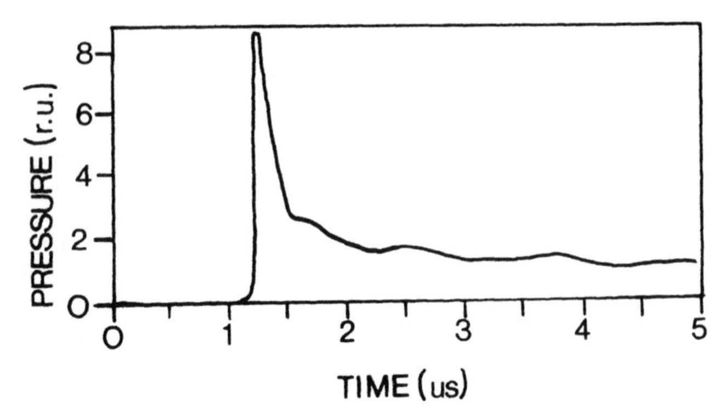

Fig.1 Shock wave shape perpendicular to the laser beam

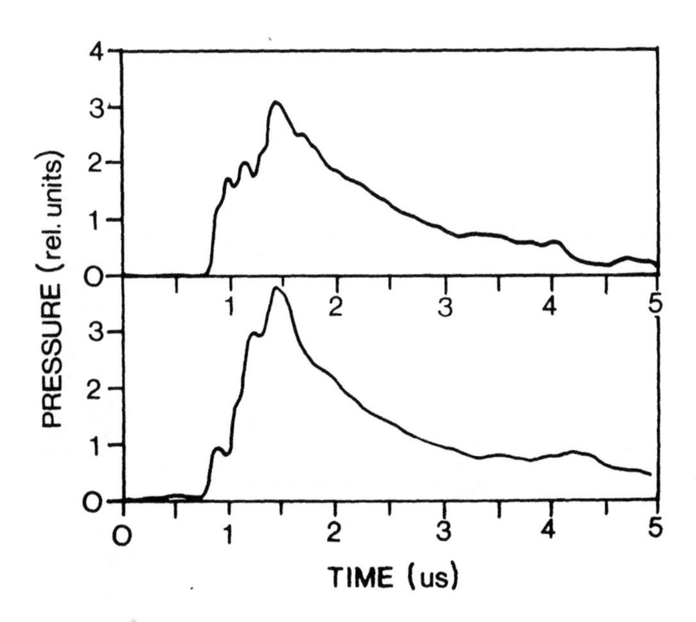

Fig.2 Shock wave shapes in laser beam direction

The rise time of a single laser induced shock wave occures in less than 3 ns, showing pressure gradients bigger than 10^{12} bar/s. On the back side the pressure decays exponentially. The shock wave duration amounts 300 to 500 ns, depending from LIB-distance and from laser energy. At a pulse energy of 60 mJ the full width at half maximum of the shock wave increases within the first 10 cm from initially 100 ns to about 300 ns.

The chain of microplasmas forms a cylindrical shock wave. Therefore the peak pressure first decreases with the square root from the distance. /2/. The bigger pressure in the direction perpendicular to the laser beam causes higher velocities than in laser beam direction. Thus after some mm a more spherical behaviour is found. Then the pressure decreases proportional 1/r. If we raise the energy the shock wave pressure increases with the square root of the laser pulse energy. /3/.

4.3. Shock Wave Energy

In order to calculate the energy W we must have knowledge of the temporal shock wave profile p(t) as well as of the absolut shockwave pressure. The more general formular given by COLE demands the shock wave velocity. /4/. But after a few mm laser induced shock waves travel with speed of sound, so that the acoustic approximation is exact enough considering the error induced by specification of the absolut shock wave pressure.

$$W = (4 \pi r^2) / (\rho_0 c) \int_0^{t_p} p^2(t) \, dt \qquad (1)$$

r is the distance from the LIB, ρ_0 is the density of the fluid, c is the speed of sound, t_p is the pulse duration. As (1) is only valid for sperical waves, we have to diminish the laser energy till single LIB occur. The shock wave energy determined under such conditions amounts to less than 1 mJ, which yields an conversion efficiency from optical to mechanical energy of about 2 per cent.

According to that relatively low shock wave energy, it becomes clear that not only energy is responsible for destrutive effects, but also shock wave duration as well as the rise time of the pressure. Because of the high damping factor of the inhomogeneous calculi materials, the shock wave mainly acts at the stone's surface breaking up tiny fragments. Moreover plasma and collapsing cavitation bubbles forming jets are in the same way responsible for stone destruction.

5. Conclusion

With help of PVDF-transducers shock wave parameters, such as shape and peak pressure as a function of distance as well as of laser pulse energy have been determined. Moreover the time resolution is high enough to measure plasma dimensions and number of shock waves generated by a single laser pulse. A prerequisite for these experimental results is an exact analysis of the generation of the electrical signal. Shock wave shape and shock wave pressure are necessary for calculation of the mechanical energy. The conversion coefficiency from optical to mechanical energy at the laser induced breakdown is only two percent.

6. Acknowledgement

This study was sponsored by the Foundation for the Promotion of Scientific Research in Austria, Project No.P4215.

7. References

1. H.Schöffmann, H.Schmidt-Kloiber, E.Reichel: J.Appl.Phys 63, Jan.1988
2. E.Reichel, H.Schmidt-Kloiber, H.Schöffmann, G.Dohr, R.Hofmann, R.Hartung: Laser in medicine and surgery 3, (1987)
3. H.Schmidt-Kloiber, E.Reichel, H.Schöffmann: Biomed. Technik, 30 (1985)
4. R.H.Cole: in Underwater Explosions, Princeton University Press, Princeton, NJ, 1948

Some Technical Aspects of Generation and Measurement of Laser Induced Shockwaves

U. Fink, Th. Meier, and R. Steiner

Institut für Lasertechnologien in der Medizin an der
Universität Ulm, Postfach 4066, D-7900 Ulm, Fed. Rep. of Germany

We present a model for a piezo-electric transducer in stripe geometry which easily allows to understand the temporal and spatial characteristics of the detector and which gives a rough estimation of the electrical signal. The calculated signal shape is compared with those from experimental recordings.

Introduction

Piezo-electric transducers are widely used for the detection and measurement of shockwaves. It has been shown that a foil from polyvinilendifluorid (PVDF) can serve as shockwave detector down to the ns-region /1/. This is gererally due to the thinness (typ. 10...20 μm) and also to small linear dimensions (1×1 mm^2).

These detectors have to be supported and contacted in a more or less elaborate manner. In contrast, we looked for a simpler design because of easier handling and, moreover, it should allow to measure the shockwaves inside a solid, e.g. a kidney stone. We found that a stripe of piezo-foil, 2×30 mm^2 large, still gave a strong electrical signal with reasonable temporal and spatial resolution.

Model

Let a shockwave be generated at the origin R_o with the initial pressure p_o. It may expand radially like a spherical shell with constant thickness and growing radius. The pressure funktion is then /1/

$$p(r) = P_o \times R_o/R. \quad (1)$$

The shockwave may hit the piezo-foil generating the voltage

$$U = Q/C, \quad (2)$$

where Q is the piezo-electrically generated charge and C the capacity of the foil. The capacity is

$$C = \epsilon_0 \, \epsilon_r \, lb/d, \tag{3}$$

where ϵ are the dielectric constants; b, c, and d are linear dimensions of the stripe, i.e. length, width, and thickness. The piezo-electrical charge Q is determined by

$$Q = d_{33} \, p \, A, \tag{4}$$

where d_{33} is the constant for the longitudinal piezo-electric effect and p is the pressure acting on the area A of the foil. This area A is not the total area l x b, but depends on the momentary radius R of the spherical shockwave. There are three phases of particular interest for the intersection of the shockwave shell which are shown in Fig. 1.

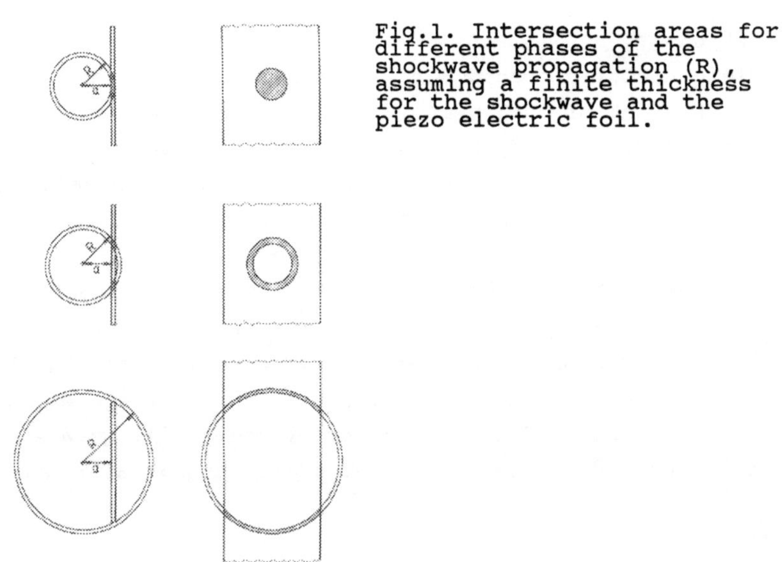

Fig.1. Intersection areas for different phases of the shockwave propagation (R), assuming a finite thickness for the shockwave and the piezo electric foil.

It turns out that for a given distance, a, between breakdown and detector foil the area, A, of the ring is nearly independend but from momentary radius R and that the absolute value is proportional to the distance a. Equations (1) to (4) allow to calculate the signal shape for different sets of parameters. Fig.2 shows some typical results. The time scale is obtained from the R-scale using normal values for the velocity of sound in water (neglecting supersonic deviations).

<u>Experiments</u>

The experimental recordings show that this rather simple model gives a satisfactory description of the signal shape. Measurements were carried out with the set-up given in another contribution of these proceedings /2/. Figure 3 shows a typical experimental signal shape.

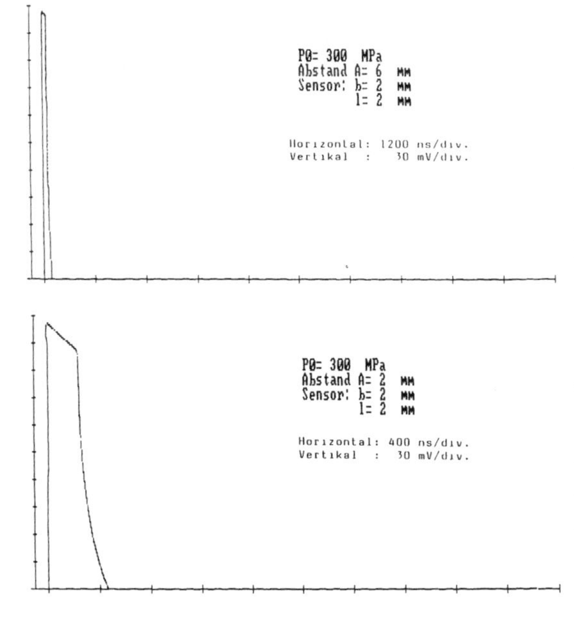

Fig. 2 a. Calculated signal shape with small detector area at distance a=6 cm. Rising part, plateau, and falling part correspond to phases shown in Fig. 1.

Fig. 2 b. Same as above, but with smaller distance a=2 mm. Signal height is nearly unchanged while duration is doubled (note the change in time scale).

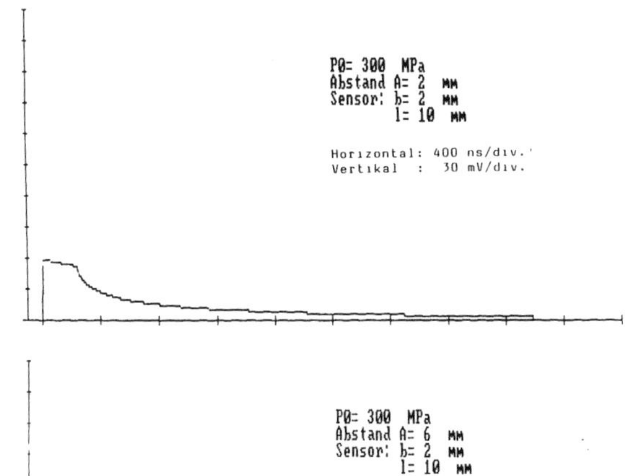

Fig. 2 c. Same as above, but with a five times longer detector. Signal height reduced to 1/5. A long tail appears.

P0= 300 MPa
Abstand A= 6 MM
Sensor: b= 2 MM
l= 10 MM
Horizontal: 1200 ns/div.
Vertikal : 30 mV/div.

Fig. 2 d. Same as above again at a larger distance a=6 mm. Good time resolution, still with tail (note the change in time scale).

The shape is, of course, smeared out compared to the computed shapes and, moreover, it may be modified by electronic effects of the amplifier. The influence of the detector length, l, is demonstrated by cutting off parts of the stripe while the other parameters are kept constant.

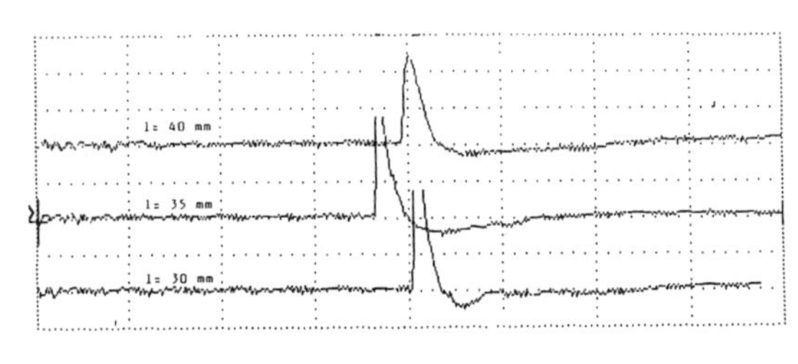

Horizontal: 500 ns/div.
Vertikal : 100 mV/div.

Fig. 3. Influence of the detector length, l, on the signal (experimental recordings with distance a = 3 mm, width of the stripe b = 2 mm)

Conclusion

The model presented helps to understand the influence of the detector geometry on the signal shape. In particular, in the case of a spherical wave and an extended plane detector a fast and strong signal is obtained as long as the wave perpendicularly hits the detector. This is a great advantage for experimental conditions when shockwaves are to be measured with only a rough knowledge of the origin or direction of the wave, e.g. in inhomogeneous matter like body concrements /2/ and other biological tissues.

References

1. H. Schmidt-Kloiber, E. Reichel, H. Schöffmann; Biomed. Technik 30 (1985) 173 - 181.

2. Th. Meier, U. Fink, R. Steiner, (this proceeding book)

The Development of an Endoscopically Applicable Optomechanical Coupler for Laser Induced Shock Wave Lithotripsy (LISL)

S. Thomas[1], *J. Pensel*[1], *W. Meyer*[1], *and F. Wondrazek*[2]

[1]Medizinisches Laserzentrum Lübeck, Peter-Monnik-Weg 9, D-2400 Lübeck, Fed. Rep. of Germany
[2]MBB Medizintechnik GmbH, Postfach 801168, D-8000 München 80, Fed. Rep. of Germany

The treatment of urinary stones has been radically changed by the introduction of extracorporeal shock wave lithotripsy. In combination with modern endo-urologic techniques 96 % of all urinary stones can be treated without open surgery. Ureteral stones, especially those impacted into the ureteral wall, represent a special therapeutic challenge for the urologist. Currently, treatment consists of ureteroscopy to the stone with ultrasonic or electrohydraulic lithotripsy. The resulting stone fragments can be extracted by wire baskets or forceps. The major set-back of this method is the need of a rigid ureteroscope for instrumental access to the stone and for visual control, necessitating anesthesia and hospitalization. Newest developments in laser medicine provide a method of destroying urinary calculi via flexible fiber systems by use of laser induced shock waves. Shockwave excitation needed to fracture calculi can be achieved in different ways.

LASER LITHOTRIPSY

CLINICAL ASPECTS — TECHNICAL USE

coupler systems	optically focussed coupler	spherically polished fiber	optomechanical coupler	bare fiber
laser-type	Q - switched Nd : YAG			Dye
LIB induction	independent	target facilitated	independet	target related
coupler - stone distance	0 - 2mm	4 - 6mm (!)	0 - 3mm	contact
fiber - coupler flexibility	(+)	++	+	+++
endoscopic use	nephroscope	ureteroscope	ureteroscope	ureteroscope
possible blind application	no (size)	no [constant distance]	yes	no [tissue damage]

MLL 87

Fig. 1: Comparison of application systems for LISL

The tunable pulsed dye laser does not require any coupling device. Initial plasma-ignition is achieved by laser-pulse energy absorption on the stone surface. The Q-switched Neodymium:YAG laser requires either a focusing device to accumulate the high power density of approximately 10 gigawatt per square centimeter needed for spontaneous dielectric breakdown in a liquid medium or an ignition device in form of an opto-mechanical coupler as first described by YANG /1/ and also by FAIR /2/. The requirements for a coupler to be used with the Q-switched Neodymium:YAG laser in clinical application can be summed up to four principles:

1) Laser induced breakdown should be originated reliably with every pulse to avoid direct tissue irradiation.

2) The power needed to destruct urinary stones of customary size should not cause any thermal effects.

3) The dimensions and the flexibility of the application system consisting of the quartz fiber, the coupler and an irrigation system must offer advantages over the currently employed techniques.

4) The durability of the coupler system must ensure that urinary stones can be destroyed reliably in one attempt without change of the application system.

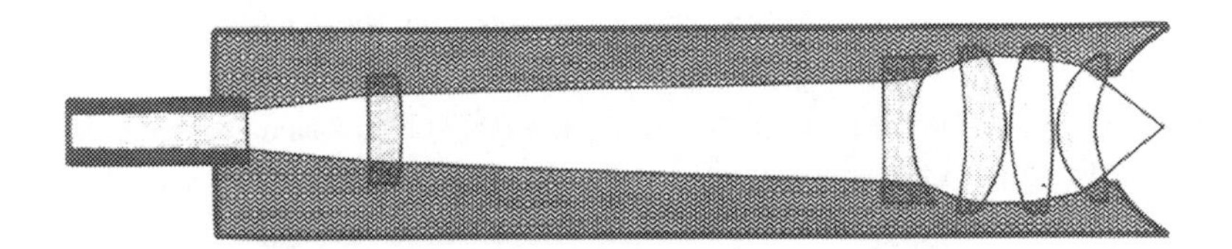

Fig. 2: Diagram of a five lens optical coupler (MBB)

Development of such a system was initiated in 1985 by MBB Medizintechnik in Ottobrunn. In vitro and in vivo assessment was performed at the Urologic Department of the Medical University Lübeck and the Medical Laser Center Lübeck. The initially developed coupler system consisted of an optically corrected focusing device of five lenses held in a brass encasing. The distal end of the coupler is formed to a spherical mirror to reflect and focus the generated shock waves to the stone. The outer dimensions of this first focusing system was 4.5 mm with a total rigid length of 3.5 cm. The energy needed to achieve dielectric breakdown is approximately 10 mJ per pulse. High repetition rates of 10 to 20 Hz are needed to guarantee rapid stone disintegration. The durability of this system was limited because evaporation processes of the brass encasing form thin film layers on the lenses leading to their destruction. The range of application is severely limited by the dimensions of this system. Clinical evaluation in two patients by way of percutaneous nephroscopy was performed showing good results, HOFSTETTER et al /3/, SCHMELLER /4/. This encouraged the development of a smaller system in combination with a 600 µm fiber. The outer diameter of this system is 2.8 mm. Laser induced breakdown is achieved with energies of about 10 mJ/pulse. This system, however, showed even less durability than the larger optical focusing system, only lasting approximately 100 pulses until one of the lenses was shattered. Major drawback of all lens-focussed systems is the fact that some form of encasing is needed to keep the lenses in place and this material is subjected to vaporization by laser energy. The resulting aerosol will cloud the lens-surface, ultimately leading to destruction of the system. Furthermore the exchange of such a system is highly complicated and can only be carried out by a technician.

Laser beam focus can also be achieved by treating the fiber-tip itself, grinding it to a highly polished spherical surface, leading to a fairly large area of focus at a distance of approximately 2-3 mm from the fiber-tip. Our own evaluation of this method shows that this could be a promising approach if the fragile fiber-tip can be preserved while guiding it through the working element of the ureteroscope to the stone and during stone-fragmentation. Furthermore the effect depends on maintaining a constant distance of 2-3 mm between fiber and stone, making constant visual control necessary.

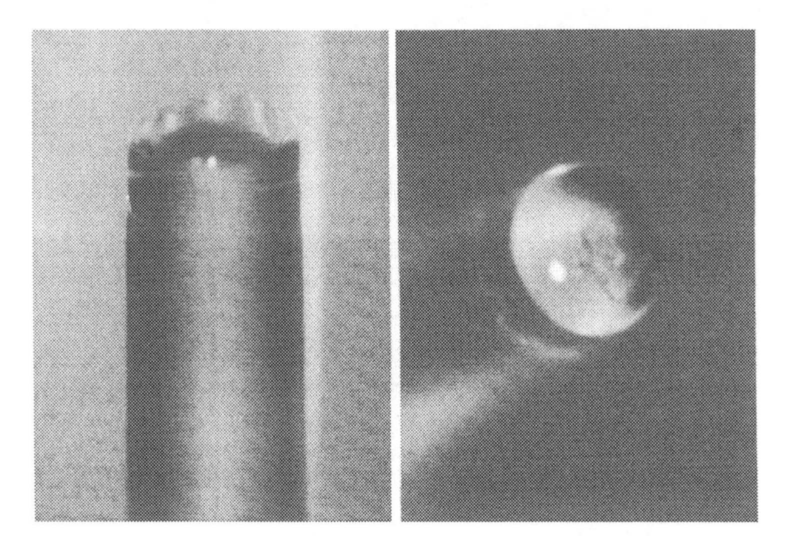

Fig.3: Spherically polished fiber tip a) before and b) after 100 laser pulses

Our efforts to reduce the dimensions of a feasible coupler have led to the construction of an opto-mechanical coupler with an outer dimension of 1.9 mm. The fluence of unfocused Q-switched YAG-laser pulses does not suffice for spontaneous laser-induced breakdown. However, materials with high linear or non-linear absorption will decrease the threshold for breakdown (WALTERS et al, /5/). This led to the construction of a metal-tip to hold a metal-bar of 0.5 mm diameter at a defined distance from the plane polished fiber-tip. The divergent laser light irradiates the metal bar. Laser pulses exceeding 20 mJ lead to ionization of the material until

irradiates the metal bar. Laser pulses exceeding 20 mJ lead to ionization of the material until sufficient electron-density is achieved to cause laser-induced breakdown. The metal material of the bar is subjected to loss.
Bar durability mostly depends on the severity and melting point of the material used. Initial experiments were carried out with a chrome-nickel steel bar. With this material only 1000 shocks were achieved before the bar broke in two. Scanning elec-

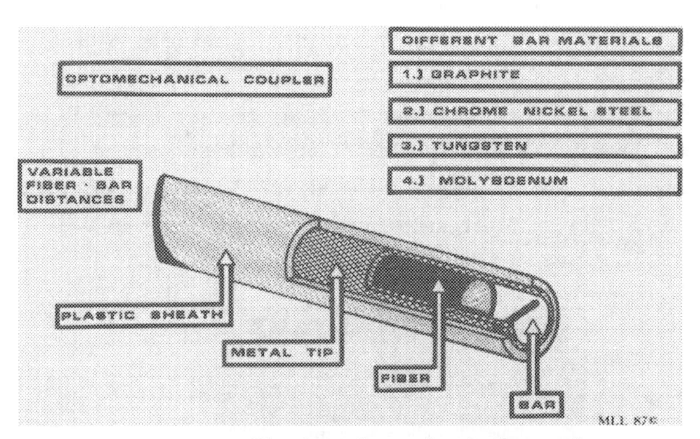

Fig.4: opto-mechanical coupler

tron microscopy of the bar surface shows distinct signs of thermal variations. Experiments at this period showed that oxalate stones of approximately 0.7 cubic cm volume require between 2000 and 4000 pulses to be broken down sufficiently. Therefore different materials of higher melting-points and higher severity were tested. Tungsten and molybdenum both have a melting-point of about 2000 degrees centigrade.

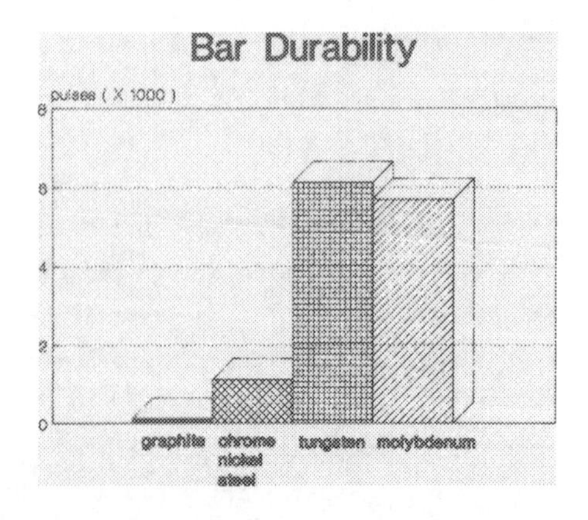

Fig. 5:
Diagram bar durability

Both metals allow over 6000 shocks before failing.
Scanning electron microscopy of the tungsten bar mainly shows cavitation effects. 85 % of the bar diameter are vaporized in laser induced breakdown; the last sixth of the bar breaks because of tungsten brittleness.
The molybdenum bar shows both thermal and cavitation effects, the whole diameter is utilized for laser induced breakdown before failing.

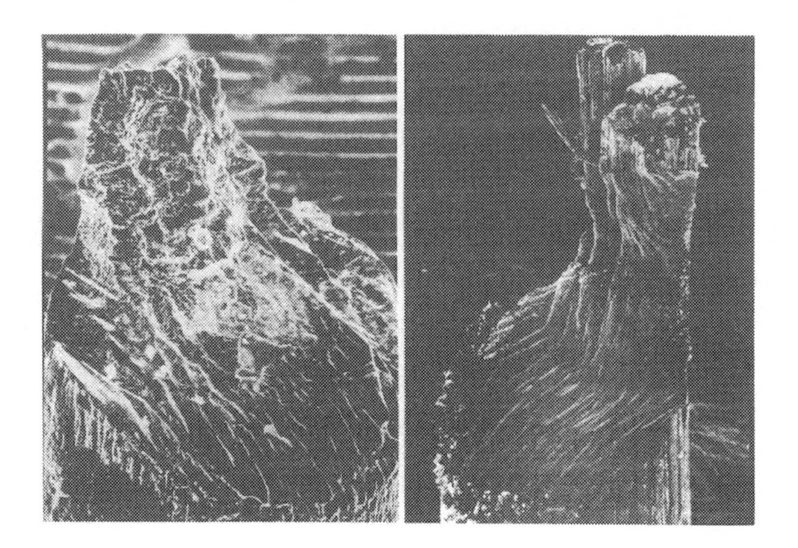

Fig.6: a) CNS BAR (1500 laser pulses) b) Tungsten bar (6800 laser pulses)

In our effort to increase the amount of achievable pulses we discovered that the fiber tip surface and not the bar is the crucial element after about 1000 shocks. If the fiber-bar distance is too small the expanding plasma wave will reach the polished fiber surface leading to cracks that will ultimately destroy the tip. Once the fiber surface is marred, energy intensity will drop rapidily and laser induced breakdown does not occur anymore.
We found the minimal fiber/bar distance to be three millimeters. Subsequent pressure measurements of the resulting shock wave at variable fiber bar distances showed the optimal efficiency and coupler durability at a fiber/bar distance of four millimeters. At this distance laser induced breakdown will occur over a range of over 6000 pulses. In the course of these shocks bar material evaporates, leading to increasing fiber/bar distances to a maximum of 4.5 millimeters. Again pressure measurements showed that shock wave pressure decreases 60 % after 6000 shocks. The

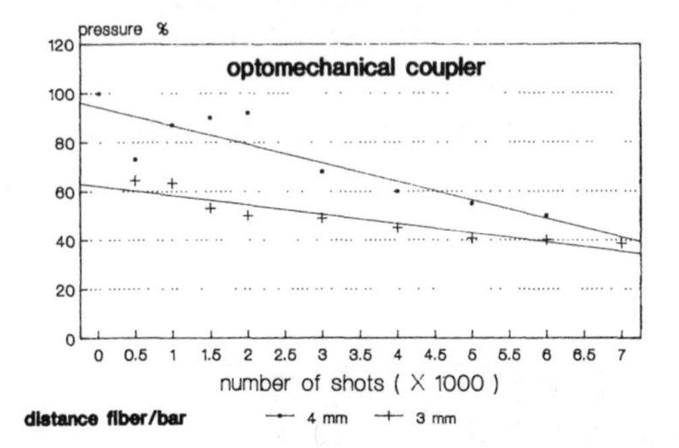

Fig. 7:

Diagram relative pressure versus amount of shock waves

pressure achieved, however, is still capable fracturing stones of all compositions. The problem that evaporated gaseous material will settle on the fiber surface leading to energy absorption and surface destruction was solved by incorporating an irrigation system in the coupler. An irrigation flow of 10 ml a minute flushes all metal and ionic debris away from the fiber surface. Although one tip is usually sufficient to complete calculus fragmentation, the optomechanical coupler we have described is easily exchangeable. If the coupler system is defective during application, the teflon irrigation tube is cut off under sterile conditions and the defective tip is removed. The fiber end can be cleaved to a perfect surface and a new tip replaced. An experienced surgeon can complete the procedure in 5 minutes. However, the low cost of the coupler calls for a one-way system that can be plugged into the laser unit at liberty.

This system is currently being developed at Medical Laser Center Lübeck. We believe that the optomechanical coupling system is the best system for laser induced shock wave lithotripsy due to its simplicity, effectiveness and tissue compatibility.

Acknowledgements :

Dr. P. Lieck for scanning electron microscope photographies and
Mr. Rene Kube for graphics and photographies.

References:

1. L.C.YANG: J. Appl. Phys. 45 (1974) p. 2601
2. H.D. FAIR: Medical Instrumentation 12 (1978) p. 100
3. A. HOFSTETTER et al.: LASER 1 (1985) p. 155
4. SCHMELLER N.T.: In Die Extracorporale und laserinduzierte Stoßwellenlithotripsie bei Harn- und Gallensteinen, ed. by M. Ziegler (Springer, Berlin-Heidelberg 1987), p. 67
5. WALTERS et al.: J. Appl. Phys. 49 , (1978), p. 2937

Current Status of Urinary Stone Therapy

K. Miller

University of Ulm, Division of Urology, Prittwitzstr. 43,
D-7900 Ulm, Fed. Rep. of Germany

Treatment of calculi in the upper urinary tract has undergone a complete change in recent years. Whereas in the period before 1980 open surgery was the only way of removing stones from the ureter or the kidney, currently open surgery represents only about 1 % of the procedures applied for the removal of urinary calculi. This dramatic change has been brought about by the development of new forms of lithotripsy. Basically two different principles are employed: contact-free lithotripsy by means of extracorporeally induced shockwaves (ESWL) and lithotripsy under endoscopic guidance using electro-hydraulic, ultrasonic and, most recently, Laser as energy source. Endoscopic guided lithotripsy is performed in the renal collecting system as percutaneous nephrolithotomy (PCNL) and in the ureter as ureteroscopy (URS).

ESWL is increasingly performed without major forms of anesthesia /10, 28/. This has been brought about by a reduction of shockwave energy and improved focussing systems. First reports /10, 28/ indicate, that a side-effect of these systems is an increasing need for repeated treatment sessions, ranging now between 20 % and 40 %.

For PCNL and URS epidural or general anesthesia is mandatory.

1. Treatment of renal calculi

Extracorporeal shockwave lithotripsy results in small stonefragments which can be discharged spontaneously. Indications for ESWL are mainly limited by the stone size (< 2 cm in diameter), as clinical studies have shown a tight correlation between stone size and postprocedural morbidity /4, 16, 17/. Moreover, results of ESWL in terms of stonefreedom after three months are significantly poorer for stones over a size of 2 cm /17/.

The insertion of a double-J ureteral stent prior to treatment has partially solved this problem as it improves the postprocedural morbidity /13/ when large stones are treated by ESWL, but there are no communicatons on improved results with this treatment combination /17/. Thus the exact limits for shockwave therapy with regard to the stone size are still to be defined and need further clinical studies /23/.

As a consequence large stones and staghorn calculi are currently considered an indication for PCNL. With this method, an appropriate part of the renal collection system is punctured under ultrasaound and fluoroscopy guidance and the transparenchymal tract is then dilated until it accomodates a 24 - 30 Fr. (0.8 - 1.0 cm) rigid endoscope. Using ultrasonic or electrohydraulic probes for lithotripsy, stones can be destroyed under vision and stone debris sucked out simultaneously or removed with grasping instruments following the process of disintegration. Comparative clinical studies have revealed PCNL equal to open surgery in terms of postoperative complications, mortality and efficacy, and superior to open surgery in terms of morbidity, recovery- and disability time and costs /1/. Nevertheless the frequency of acute complications following PCNL for staghorn-stone removal is high: fever 55 - 60 % /2/, hemorrhage requiring blood transfusion 43 % /27/ or even nephrectomy 1 - 5 %, perirenal extravasation 25 % /27/ and pleural reaction 18 % /27/. Late sequelae encompass strictures of the UPJ in 3 % /28/ and delayed bleeding from a.-v.-fistula in 3 %. According to clinical studies /19/ renal function is in general not compromised by PCNL.

To accomodate the rigid ultrasonic probe, rigid nephroscopes are necessary. Thus, depending on the individual anatomic situation, only certain parts of the branched renal collecting system can be reached via one access, resulting often in incomplete stone removal. To overcome this problem, different approaches can be persued:

- Additonal percutaneous tracks are installed in the same session.
- Residual stone fragments are treated by ESWL in additional sessions.

The former approach bears an increased risk of complications, the latter is more inconvenient for the patient, as repeated treatment sessions must be expected.

The use of a flexible nephroscope could solve these problems, but with the comparably voluminous electrohydraulic probes it could not be combined without a the side effect of a reduced irrigation flow causing poor vision. Laser-fibres of only 200 um diameter, which have already proved their efficacy for lithotripsy of ureteral calculi /5/, could therefore give a great impact on flexible nephroscopy and nephrolithotripsy, expectedly leading not only to complete stone removal in one session but also to a lower complication rate.

2. Treatment of ureteral calculi

As with renal calculi, extracorporeal shockwave-lithotripsy has rapidly replaced surgical procedures for the treatment of ureteric calculi. In the beginning, indications for ESWL have been confined to stones in the ureter above the pelvic brim /2, 7, 21/, as the lower ureter has been considered to be not accessible for shockwaves on account of the overlying bony plevis. This problem has been overcome by a modification of the positioning of the patient on the lithotriptor stretcher /22/, thus adding lower ureteric calculi to the indications of non-invasive stone management.

2.1 Concepts for upper ureteral calculi

Early experience with in-situ shockwave lithotripsy of ureteric calculi has revealed a lower disintegration rate, (60 - 75 %) /7, 21, 25/, when compared with kidney stones (>90 %) /2, 7/. Experimental studies indicate, that the lack of fluid between the stone and the ureteric wall impairs the efficacy of shock-waves /25/. As a consequence, different strategies for the management of calculi in the upper ureter have emerged /3, 6/:

A. Retrograde manipulation of the stone prior to ESWL:

When the stone is pushed or flushed back into the kidney by means of a ureteric catheter, subsequent disintegration is easily achieved. Unfortunately, the retrograde manipulation results only in 40 - 60 % in an efficient mobilisation of the calculus /9, 14/. Thus about 50 % of the stones must be treated in-situ after failed mobilisation. Moreover perforation of the ureter occurs in up to 5% in conjunction with the retrograde manipulation /15/. Nevertheless, this approach is persued in many centers as it offers the best chance of a successful treatment in one session /8, 14, 15, 25/.

B. Shockwave treatment in-situ:

The rational for treatment in-situ is to avoid any invasive procedures and rather to take the higher risk of an additional ESWL session, when no complete stone disintegration has been achieved. This approach results in a successrate of 80 - 85 % including 10 - 14 % second sessions /9, 14, 15/. With anesthesia-free treatment lately beeing available, in-situ treatment may become more popular, as retrograde manipulations without anesthesia cause considerable discomfort, particularly in male patients.

2.2. Concepts for distal ureteric calculi

For the treatment of calculi in the pelvic part of the ureter some adjustments on patient positioning are mandatory. Thus, direct fluoroscopic stone visualization is possible in about 70 % of the patients, whereas in 30 % contrast application is mandatory for exact identification of the course of the ureter /22/.

Treatment of distal ureteric calculi in-situ is successful in 93 %, with the need of more then one session in 14 % of the cases /22/. Extracorporeal shock-wave-lithotripsy offers marked advantages, when compared with endourological procedures: using ureteroscopy (URS) for stone removal, damaging of the ureter occurs in about 10 % /12, 18, 20/, whereas no complications have yet been encountered after ESWL /22, 24, 26/. Shockwave treatment can be performed under opioid analgesia, whereas general or epidural anesthesia is neccessary for uretersocopy. Moreover URS can be rather time consuming, particularly when ultrasonic lithotripsy is required prior to stone removal /18/. These advantages are promoting ESWL to the treatment of choice for distal ureteric calculi in an increasing number of centers.

Endoureteral Laser-lithotripy, using either the pulsed dye or the Neodym-Yag laser /5, 11/, can currently be rated as an alternative to endoscopically guided ultrasonic lithotripsy using conventional rigid ureteroscopes. Initial clinical experience indicates, that the laser is efficient and safe /5, 11/. However, to compete with ESWL as the method of choice for the management of ureteral calculi, laser-lithotripsy must be performed anesthesia-free, i.e. with flexible ureteroscopes or under fluoroscopic guidance. To assess the feasibility of this, further experimental and clinical studies will be necessary.

3. Literature

1. Brannen G.E., Bush W.H., Correa R.J., Gibbons R.P., Elder J.S.: Kidney stone removal: percutaneous versus surgical lithotomy. J. Urol. 133: 6 (1985).
2. Chaussy Ch., Schmiedt E., Jocham D., Brendel W., Forssmann B., Walther V.: First clinical experience with extracorporeally induced destruction of kidneystones by shockwaves. J. Urol. 127: 417 (1982).
3. Chaussy C., Fuchs G.: ESWL: what is proven, what is controversial. IV. World Congress on Endourology and ESWL, Madrid (1986).

4. Drach G.W.: Report of the United States Cooperative study of Extracorporeal shockwave lithotripsy. J. Urol. 135: 1127 (1986).
5. Dretler S.P., Watson G., Parrish J.A., Murray S.: Pulsed Dye Laser Fragmentation of Ureteral Calculi: Initial Clinical Experience. J. Urol. 137: 386 (1987).
6. Eisenberger F., Rassweiler J.: Upper ureteral calculi - push or bang? IV. World Congress on Endourology and ESWL, Madrid (1986).
7. Fuchs G., Miller K., Raßweiler J., Eisenberger, F.: One year experience with the Dornier Lithotripter. Eur. Urol. 11: 145 (1985).
8. Fuchs G., Lupu A.N., Chaussy C.: Treatment of ureteral stones: controversies and current differential indications. IV. World Congress on Endourology and ESWL, Madrid (1986).
9. Graff J., Pastor J., Mach P., Michel W., Funke P.J., Senge Th.: Extra-corporeal shockwave (ESWL)- treatment of ureteral stones - an analysis of 417 cases. J. Urol. 137: 143A, (1987).
10. Graff J., Pastor J., Herberhold D., Hankemeier U., Senge Th.: Technical modifications of the Dornier HM-3 lithotripter with an improved anesthesia technique. World J. Urol. 5: 202 (1987).
11. Hofmann R., Hartung R., Geißdörfer K., Ascherl R., Erhardt W., Schmidt-Kloiber H., Reichel E.: Laser Induced Shock Wave Lithotripsy (LISL) - Biologic Effects and First Clinical Application. Laser Med. Surg. 3: 247 (1987).
12. Huffmann J.L., Bagley D.H., Schoenberg H.W. Lyon E.S.: Transurethral removal of large ureteral and renal pelvic calculi using ureteroscopic ultrasonic lithotripsy. J. Urol. 130: 31, (1983).
13. Libby J., Griffith D.G.: Large calculi and ESWL: is morbidity minimized by ureteral stents? J. Urol. 135: 182 (1986).
14. Jenkins A.: ESWL treatment of ureteral calculi. J. Urol. 135: 182A (1986).
15. Lingeman J.E., Newman D.M., Mertz H.H.O., Mosbaugh Ph.G., Steele R.E., Knapp P.M., Shirell W.: Management of upper ureteral calculi with ESWL. IV. World Congress on Endourology and ESWL, Madrid (1986).

16. Lingeman J.E.: Extracorporeal shockwave lithotripsy - the Methodist Hospital experience. J. Urol. 135: 1134 (1986).
17. Lingeman J.E.: Current concepts on the relative efficacy of percutaneous nephrostolithotomy and extracorporeal shockwave lithotripsy. World J. Urol. 5: 229 (1987).
18. Lyon E.S., Huffman J.L., Bagley D.H.: Ureteroscopy and ureteropyeloscopy. Urology 23: 29, (1984).
19. Mayo M.E., Krieger J.N., Rudd T.G.: Effect of percutaneous nephrostolithotomy on renal function. J. Urol. 133: 167 (1985).
20. Miller K., Gumpinger R., Fuchs G., Rassweiler J., Eisenberger F.: 160 cases of ureteroscopy. J. Urol. 133/2: 171 (1985).
21. Miller K., Fuchs G., Rassweiler J., Eisenberger F.: Treatment of ureteral stone disease: the role of ESWL and endourology. World J.Urol. 3: 53, (1985).
22. Miller K., Bubeck J.R., Hautmann R.: Extracorporeal shockwave lithotripsy of distal ureteral calculi. Eur. Urol. 12: 305, (1986).
23. Miller K., Bachor R., Hautmann R.: Percutaneous nephrolithotomy/ESWL versus ureteral stent/ESWL for the treamtent of large renal calculi and staghorn stones: a prospective randomized study. V. World Congress on Endourology and ESWL, Cairo (1987).
24. Miller K., Hautmann R.: Treatment of distal ureteral calculi with ESWL: experience with more than 100 consecutive cases. World J. Urol. 5: 259 (1987).
25. Müller S.C., v. Haverbeke J., El Seweifi A., Alken P.: Der hohe Harnleiterstein - ein Problem trotz extrakorporaler Stoßwellenlithotripsie. Akt. Urol. 16: 294 (1985).
26. Rassweiler J., Hath U., Lutz K., Eisenberger F.: In-situ ESWL beim tiefen Harnleiterstein. Akt. Urol. 17: 328, (1986).
27. Snyder J.A., Rosenblum J.L., Smith A.D.: Endourological removal of staghorn calculi in the elderly: analysis of 42 cases. J. Endourol. 1: 123 (1987).

28. Zwergel U., Neisius D., Zwergel T., Ziegler M.: Results and clinical management of extracorporeal piezoelectric lithotripsy (EPL) in 1321 consecutive treatments. World J. Urol. 5: 213 (1987).

The Clinical Experience With Pulsed Dye Laser Photofragmentation of Ureteral Calculi (127 Cases)

S.P. Dretler

Lithotriptor Unit, Massachusetts General Hospital Boston,
Boston, MA 02114, USA and
Associate Professor in Urology, Harvard Medical School

Since September of 1985 until present, we have used the pulsed dye laser for transureteroscopic photofragmentation of ureteral calculi. /1,2/. Recently, we have also used the same laser percutaneously via a flexible nephroscope to fragment calculi sequestered in calyces that are unable to be reached by the percutaneous rigid ultrasonic sonotrode. Herein is presented a review of the instrumentation and techniques developed for percutaneous and ureteral laser lithotripsy, the results of laser fragmentation of 127 ureteral calculi and a discussion of the principles and possibilities of laser lithotripsy.

Materials and Methods

One hundred and twenty-seven patients with ureteral calculi too large to be extracted by basket were selected for treatment by transureteroscopic laser methods with the MDL Scientific Model Laser Lithotriptor.* Included were 12 patients with UPJ or upper ureteral calculi (L1 -L3) which were impacted, unable to be bypassed by ureteral stent or had failed previous ESWL. Twenty-two mid ureteral calculi (L3 - S1) were treated. Initially, only patients with mid ureteral calculi unable to be bypassed by stent prior to ESWL were included; after increased experience with laser lithotripsy, patients with mid ureteral calculi were selected for laser therapy if the stone was impacted and would require manipulation prior to ESWL or if the stone was not densely calcified ("radio-fragile"), it was anticipated that it may be difficult to localize by ESWL and appeared easily fragmentable by laser lithotripsy. Eighty-two lower ureteral calculi too big to be basket extracted were treated as well as 7 patients with calculi impacted in the intramural tunnel and 4 patients with lower ureteral steinstrasse (Figure 1).

---

* Candela Laser Corporation, Wayland, MA 01778

Additionally, two patients with cayceal calculi were treated by laser lithotripsy via a percutaneous flexible nephroscope.

The technique used for laser fragmentation utilized a 250 micron silica-coated quartz fiber. The same fiber was used for both ureteral and percutaneous laser lithotripsy. The laser fiber was used within a 4 F ureteral catheter for direct application (Figure 2) or was passed into a flat-wire stone basket via a hole machined through the center of the wire (Figure 3)/2/. The laser in the catheter and in the basket have been used through the working channel of the 9.5 F Wolf rigid ureteroscope. Additionally, 7 F and 9 F non-steerable flexible ureteroscopes have been used as has a newly developed prototype 7.2 F semi-rigid ureteroscope.

Rigid ureteroscopy was performed by the following method: cone tip retrograde ureterography was carried out under C-arm fluoroscopic control. After ureterography, an .038 wire was passed by the stone to the renal pelvis and left in place throughout the procedure. If the stone could not be bypassed with a wire, the wire was passed to an area just below the calculus and left in place. For calculi lodged within the intramural tunnel, too distal to allow effective wire placement, treatment was carried out without a guide wire in place.

If the ureteral orifice was small or if the ureter below the stone was too narrow to accept the 9.5 F rigid ureteroscope, then ureteral dilatation with a 6 mm ureteral dilatation balloon was performed. If an indwelling ureteral stent had been in position, then ureteral balloon dilatation was performed only if attempts to pass the ureteroscope met with resistance. For calculi lodged too distal to allow 6 mm balloon dilatation, a dilating balloon which could be passed through the ureteroscope and inflated to 12 F, was used to dilate the meatus and distal tunnel. Once dilatation was carried out, the 9.5 F rigid ureteroscope was introduced to the level of the calculus. If the calculus looked fragile (struvite, uric acid or calcium oxalate dihydrate), the laser fiber (inside a 4 F ureteral catheter) was placed directly on the calculus and the stone was fragmented until all pieces were 1 - 2 mm in size. If, when first visualizing the stone through the ureteroscope, it looked like the more difficult to fragment calcium oxalate monohydrate(black, nodular) or if initial fragmentation of the fragile outer layer revealed a black calcium oxalate monohydrate core the stone was

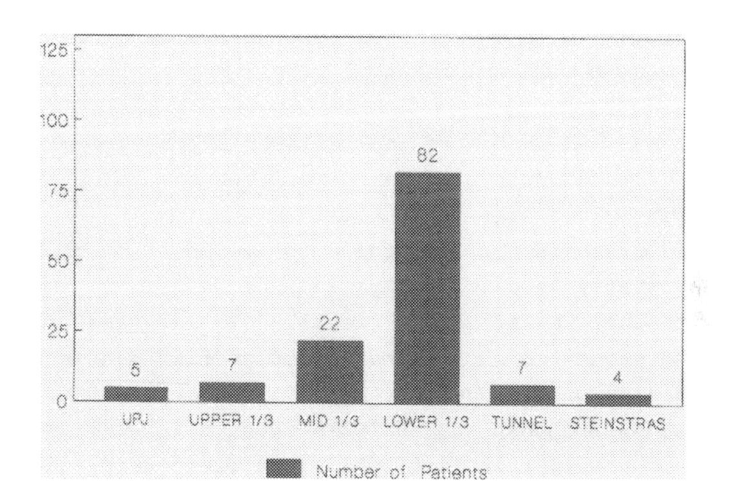

Figure 1 Level of ureteral stones treated by laser lithotripsy

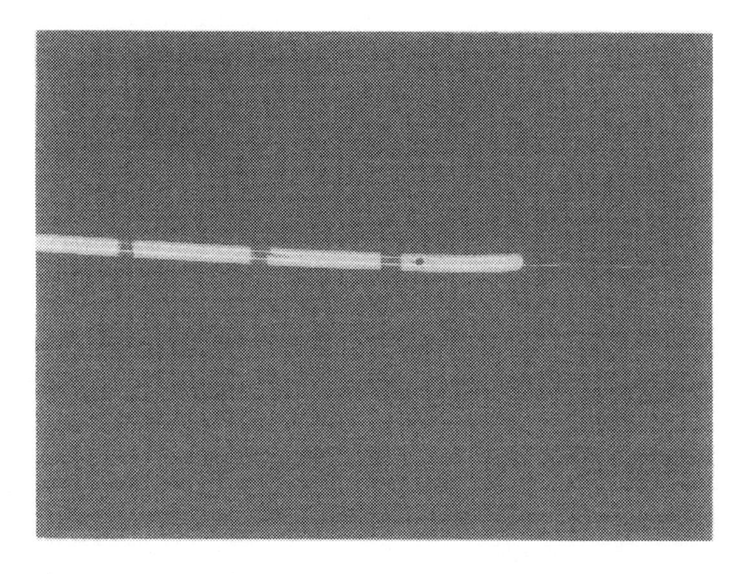

Figure 2 250 micron laser fiber inside a 4 F ureteral catheter

entrapped in the "laser basket" and held tightly during laser application. This technique maximized the effect of the laser induced shock wave. Laser fragmentation was then carried out until the diameter of the stone became small enough so that the core could be extracted with the laser basket.

The 9.5 F rigid ureteroscope and laser fiber were also used to disimpact UPJ and upper ureteral calculi unable to be stent bypassed or dislodged by catheter manipulation or irrigation techniques. The 9.5 F rigid ureteroscope was passed to the point of obstruction. The stone was usually seen to be impacted into a narrow, swollen ring of ureter with only the stone tip visualized. The laser was placed on the exposed stone tip and fragmentation carried out until the stone could be disimpacted (Figures 4A & 4B). At this point, the large residual stone fragment was manipulated back into the renal pelvis, a stent placed and ESWL carried out. Less frequently, the ureteroscope was advanced through the swollen area of the ureter and upper ureteral stone fragmentation completed in-situ. If the stone was seen to be composed of the harder to fragment calcium oxalate dihydrate, then attempts were made to complete fragmentation in-situ without the need for ESWL. In two patients, the stone was fragmented above a narrow or strictured area of ureter which could not be passed with the ureteroscope. On these occassions, the "laser basket" was passed up beyond the ureteral narrowing and the stone entrapped in the basket. The laser fiber was passed up the laser basket and blind fragmentation carried out until the basket and its smaller fragments could be extracted through the narrow or tortuous ureter. The above summarizes the methods used via the 9.5 F rigid ureteroscope.

The laser fiber was also used inside the 7 F and 9 F flexible, non-steerable Reichert ureteroscopes. This technique was attempted 18 times: when the ureter was too narrow to accept the rigid ureteroscope; when the calculus was impacted in the intramural tunnel; when a UPJ stone required disimpaction; as an attempt to avoid the necessity for ureteral balloon dilatation; and, for mid ureteral calculi that looked fragile on abdominal flat film (radio-fragile) and would not need "laser basket".

The newly developed 7.2 F semi-rigid ureteroscope was used in 7 patients. This 40 cm ureteroscope uses flexible optical fibers so that it may be bent without loosing vision yet retains the torque and

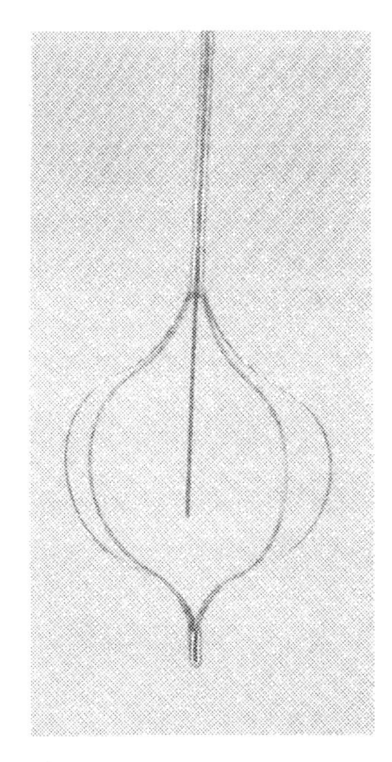

Figure 3 Laserbasket: laser fiber enters the center of the wire; hard stone is held by basket against fiber

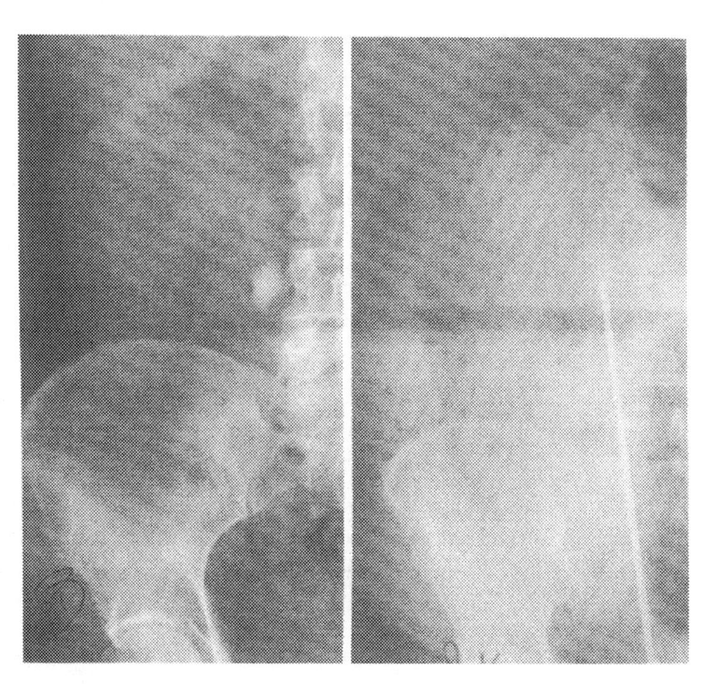

Figure 4A Stone impacted at ureteropelvic junction

Figure 4B Laser induced partial fragmentation and disimpaction

maneuverability of a rigid ureteroscope. The distal 15 cm is 7 F and then it gradually enlarges so that the proximal portion (the area used as a fulcrum) is 11 F. It has two channels, each .028 cm, one for irrigation and the other for the laser. This instrument has been used for impacted distal ureteral calculi. The 7 F semi-rigid ureteroscope is able to be advanced directly to the distal ureteral calculus without the need for ureteral dilatation.

The MDL scientific model pulsed dye laser used a coumarin green dye and gererated a wavelength of 504 nm an a pulse duration of one microsecond. In the scientific model, the laser fiber had to be alligned with the laser light output for each use. The repetition rate of laser pulsations was 5 Hz or 10 Hz and the energy used was 60 mJ per pulse. The MDL was located at a remote site and the 250 micron silica coated quartz optical fiber was run in the ceiling of the operating suites, one to the cystocopic area for ureteroscopic use and one to the operating room for percutaneous use. The laser power output was calibrated for each procedure. After the first 25 procedures, which used pulse levels varying from 20 - 60 mJ/pulse, the power was kept at 60 mJ per pulse. If, during the procedure, the laser fiber came out of alignment and could not be realigned in a reasonable period of time or, if a mechanical problem in the flashlamp laser occurred during the procedure or adequate power (60 mJ/pulse) could not be maintained during the operation, the laser portion of the procedure was terminated and the case was classified as a "machine failure". If, despite the use for 60 mJ/pulse, the stone would not fragment, the case was classified as a "laser failure". Stone fragmentation was "total" if all fragments were less than 3 mm and able to readily pass. Fragmentation was partial if the laser basket was required to complete stone removal.

Post-operatively, ureteral stents were used routinely during the initial experience with the laser, especially if ureteral balloon dilatation had been performed. With further laser experience, stents were omitted as often as appeared safe. Now, with the availability of indwelling stents with "pull-out strings", stenting is again done routinely. The number of stents used says nothing about ureteral trauma or laser action.

Laser induced ureteral injury including purpura, abrasion or perforation was noted for each patient. Abdominal radiographs were routinely obtained immediately after treatment on all patients to assess size and number of residual stone fragments. All patients were followed until fully recovered from their procedure and were free of symptoms and the offending stone.

Each ureteral stone treated by laser had its area calculated by multiplying by length an width and recorded as sq. mm. The number of laser pulses for each stone was recorded. The stones were classified according to stone composition. A calculation was made for the number of laser pulses required per sq. mm of stone of each composition.

Two patients underwent laser fragmentation by the percutaneous route. In both instances, percutaneous ultrasonic lithotripsy had been performed and inaccessible fragments were seen to be left behind. Each patient was a poor candidate for ESWL. A flexible nephroscope was used in each instance. The laser fiber was used inside a 4F ureteral catheter and passed through the working channel of the nephroscope.

<u>Results</u>

One hundred and twenty-seven patients had laser treatment or attempted laser treatment of calculi within the ureter. The smallest stone was 5 x 5 mm and the largest measured greater than 10 x 25 mm. In 123 patients, laser was successful in totally fragmenting the calculus to spontaneously passable particles smaller than 3 mm in 73 patients (59.3%). In 34 patients (27.6%), partial fragmentation was carried out and the remainder of the stone was extracted by basket. There were 4 patient with steinstrasse. There was inadequate power delivery to the tip of the laser fiber on 3 occasions and those patients were termed "machine failures". This latter circumstance has not occurred in the last 75 cases. In 8 patients (6.4%), laser fragmentation partially or minimally fragmented the calculus and because of the hardness of the stone (monohydrate, brushite, cystine) or the presence of steinstrasse, the completion of therapy required ultrasonic lithotripsy. In 9 patients, the laser was used to dis-impact the stone and ESWL used as definitive therapy.

In 24 of 127 patients, the guide wire was unable to bypass the stone. Laser fragmentation, with or without basket extraction, was used successfully in 21 of these 24 patients. Supplemental ESWL was required for 3 patients. In 91 patients, there was no ill effect of any action by the laser on the ureteral wall. Purpura, erythema and bleeding occurred in 29 instances. However, to my observation, this was due to the mechanical action of the stiff 250 micron quartz fiber, not to any observable laser action. If the laser was discharged on the ureteral wall, a 250 micron (barely visible) mucosal burn occurred. Partial tear of the ureter occurred at the site of stone impaction in 6 instances and in each it appeared to be due to the mechanical effect of forceful advancement of the ureteroscope through the swollen ureter rather than laser action. The sharp quartz fiber was mechanically pushed through the ureteral wall a number of times and the laser energy inadvertently discharged. There were no adverse effects; the hole produced in the ureter, being only 250 microns wide, could barely be seen and there was no evidence of laser damage to adjacent organs or to ureteral wall. In one patient, the effect of the laser on the ureteral wall was not recorded.

The flexible 7 & 9 F ureteroscopes (non-steerable) were successfully used for fragmentation in only 8 of 18 attempts. The failures were related to non-steerability and the inability to direct the laser fiber to the stone.

The semi-rigid ureteroscope was successful in 4 of 7 patients, all females with impacted distal calculi. The ureteroscope could be easily advanced into the distal ureter in females but the prototype was difficult to use in male patients.

There was no complication related to laser action. Two patients (the first two in the series) developed ureteral strictures at the site of ureteral dilatation and have been previously reported /2/. Two patients have developed narrowing of the ureter at the site of previous stone impaction without proximal ureteral or calyceal dilatation. Two patients required percutaneous nephrostomy to treat temporary obstruction at the site of stone treatment. One patient required a second ureteroscopy because of failure to reach the stone on the first attempt. No patient has required a second treatment because of remaining stone fragments. One patient had a residual 2 mm fragment which was pushed through the ureteral wall and left in a false passage. There have been no sequelae. One patient had a 3 mm fragment noted at initial follow up KUB and spontaneously passed this fragment within one month. No other patient had a residual fragment greater than 2 mm in size. Except for the strictures already described, there have been no other long term effects of this form of therapy. No patient who has undergone laser lithotripsy has required ureterolithotomy.

Percutaneous treatment was successful in the two instances used. Both calculi were composed of struvite and were easily fragmented.

The fiber was directed through the flexible nephroscope and laser action did not result in mucosal abrasion or bleeding. In no instance of ureteroscopic or percutaneous laser lithotripsy did the quartz fiber break within the patient.

Discussion

The pulsed dye laser ist used with a fluorescent green dye which emits at a wavelenght of 504 nm. This light wavelength was chosen because it is readily absorbed by the yellow or black pigment of the stones and not by the hemoglobin of surrounding tissue. This selective absorption is the basis for the ability of the laser to fragment calculi and not affect the adjacent ureteral wall /3/. Previous attempts at laser induced stone destruction with Nd-YAG lasers caused stone melting and with the melting, adjacent tissue injury. Because the Nd-YAG wavelength (1064 nm) is poorly absorbed by the stone, enormous energies were necessary for stone destruction, equivalent to 10,000 x the 60 mJ/pulse used by the pulsed dye laser. Similarly, it should be noted that electrohydraulic lithotripsy (EHL) produces 80 x the energy of the laser (5 joules vs 60 mJ). The lack of selective stone absorption and the 80 fold increase in energy level of EHL probably accounts for the commonly observed adjacent tissue damage caused by EHL and the lack of this injury during laser lithotripsy. The success of the pulsed dye laser over prior continuous laser radiation methods is a result, not only of absorption of light by the stone, but also of the short one micro-second duration and minimal heat production of each pulse. Although there is a microscopic focus of heat produced at the point of light absorption by the stone pigment, bulk heating of the stone does not occur. The proposed mechanism of stone destruction is that each pulse of laser light is absorbed by the stone, causing a small "plasma" of ions and electrons to occur. This plasma, or small gas bubble, then absorbs further laser irradation and expands, causing a fracture lines to occur in the stone /4/. The use of saline irrigation during stone fragmentation confines the plasma and the shock wave and facilitates fragmentation. As with ESWL, fragmentation rates are related to the fragility of the calculus. Calculi composed of calcium oxalate dihydrate, struvite and uric acid are very fragile and laser fragmentation may be carried out in 1-2 minutes or less, the shortest times for stone fragmentation being 5-10 seconds. However, the calcium oxalate monohydrate, brushite and cystine calculi are resistant to fragmentation and must be chipped down and then the core is basket extracted when small enough to remove. The amount of laser energy needed to complete fragmentation is much higher for a CaOx monohydrate than for a calcium oxalate dihydrate stone. This is an experience similar to that noted when stones of mono and dihydrate were treated in the lithotriptor.

For effective laser stone fragmentation, the laser fiber must be placed directly on the calculus. If the fiber does not touch the stone, fragmentation will not occur. There can be no blood clot between the tip of the fiber and the stone or the laser energy will not be absorbed and a plasma and subsequent shock wave will not be formed. Saline irrigation must be used during the procedure.

As laser energy is absorbed by the calculus, there is an audible acoustic report, sounding like "tic, tic, tic...", with each pulsation. If, during treatment, the quartz fiber slides by the stone, is pushed into the ureteral wall or is not in contact with the stone, the audible report ceases. The fiber must again be placed in contact with the stone. The fiber may be pushed back inside the ureteral catheter during treatment. In that case, the laser will not work. If fragmentation ist not progressing, these small problems may be the cause.

When hard stones are encountered, the laser must be directed at what seems to be a crevice or a week area of the stone and energy must be applied to this area until fragmentation occurs. This is the method by which hard stones can be chipped away and then the core can be basket extracted. Currently, a disadvantage of the laser is that it is difficult to break calcium oxalate monohydrate calculi. Recently, to solve this problem, Murray did studies on the effect of a larger (320 micron core diameter vs 250 core diameter) fiber on the fragmentation of the harder, calcium oxalate monohydrate calculi and found that the larger fiber improves the fragmentation rate, without the need for increasing the energy /5/.

The laser has been used successfully (4/4) to disimpact steinstrasse. Its advantage is that it can be advanced into the packed ureteral orifice and disimpact the calculi without causing adjacent wall injury and without the need for prior ureteral dilation. Once the distal few cms of steinstrasse are disimpacted, the ultrasonic probe can then be inserted in the dilated distal ureter and steinstrasse removal can be completed.

Percutaneous use of the laser is a natural outgrowth of laser ureteral lithotripsy. The fiber can be bent to a 70 degree angle and may be advanced into and used in narrow necked calyces without causing tissue injury or bleeding frequently seen with EHL. The laser is a valuable addition to the percuaneous armamenterium and should reduce the need for ESWL of residual sequestered calyceal calculi after percutaneous debulking procedures.

The results of two years of laser ureteral lithotripsy are encouraging and indicate a definite role for this modality in the urologic armamenterium. We have used it for all lower ureteral stones, fragile midureteral stones, impacted ureteral stones at all levels and percutaneously. Although many modalities (ESWL, EHL, Ultrasonic) are used for ureteral calculi, there can be little argument that for the impacted lower ureteral stone, for initiating the treatment of steinstrasse and for calculi impacted in ureters which can only be traversed by small flexible ureteroscopes, the size, precision, safety and effectiveness of the laser makes it the modality of choice. Whether the indications expand to justify widespread use will depend on the ability of the laser to fragment calcium oxalate monohydrate calculi and the development of supporting micro-instrumentation to facilitate use within the upper urinary tract.

Laser stone fragmentation was initially designed for use within the lower ureter for those calculi not able to be treated by ESWL. However, recent reports of hypertension occurring in 8% of patients after ESWL /6/ suggest that transureteral laser lithotripsy may have a wider application than initially anticipated. The use of laser fibers in small caliber flexible steerable ureteroscopes which will reach into the renal pelvis and calyces promises to offer an alternative to ESWL in selected patients.

REFERENCES

1. Dretler, S.P., Watson, G., Parrish, J.A., Murray. S: Pulsed Dye Laser Fragmentation of Ureteral Calculi: Initial Clinical Experience. J. Urol, 137:386, 1987

2. Dretler, S.P.: Laser Photofragmentation of Ureteral Calculi: Analysis of 75 Cases. J. Endourology, Vol 1, Number 1, pg 9, 1987

3. Watson, G., Jacques, S.L., Dretler, S.P. Parrish, J.A.: Tunable Pulsed Dye Laser for Fragmentation of Urinary Calculi. Lasers Surg Med, 5:160; 1985

4. Nishioka, N.: Mechanism of Laser Induced Stone Fragmentation of Urinary and Biliary Calculi. Lasers in the Life Sciences, L(3):231-245, 1987

5. Murray, S. (Candela Laser Corporation): Personal Communication, 1987

6. Lingeman, J.E., Kulb, T.B.: Hypertension Following Extracorporeal Shock Wave Lithotripsy. J. Urol, 137:142A, 1987

Clincial Experience with Laser-Induced Shock-Wave Lithotripsy

R. Hofmann[1], *R. Hartung*[1], *H. Schmidt-Kloiber*[2], *E. Reichel*[2], *and H. Schöffmann*[2]

[1]Department of Urology, Technische Universität München, Klinikum rechts der Isar, D-8000 München 80, Fed. Rep. of Germany
[2]Department of Experimental Physics, Karl-Franzens-University, A-8010 Graz, Austria

Extracorporeal shockwave lithotripsy(ESWL)has changed management of urinary calculi considerably to a non invasive procedure.Nevertheless ureteral stones,especially those being impacted for a longer period in the ureter or very hard ones(e.g. calcium-oxalate-monohydrate or uric acid stones)are still a problem.Ureteral calculi comprise of up to 50% of all stones now and can cause considerable discomfort for the patient.Ureteroscopic manipulation with rigid instruments so far was only possible with rigid ultrasonic probes,baskets or forceps.
The aim of our research was to develop an effective,secure and minimal invasive method for disintegration of urinary calculi.Thin flexible probes,which can be brought to the stone by small,preferably active flexible endoscopes should be used.High intensity laser energy,which can be transmitted through small quartz fibres seemed to be suitable.

Continuous wave lasers and pulsed lasers up to the microsecond range induce thermic lesions in an irradiated tissue by transformation of laser energy into heat.The applied laser light results in deep coagulation effects with damage of the adjacent urothelium.Urinary stones cannotbe disintegrated by continuous wave lasers without thermic tissue damage(1,2).Laser energy from a pulsed Neodymium YAG laser(Nd-YAG) with a pulse duration in the nanosecond range however is changed into mechanic energy as shock waves by creation of a localized plasma.This procedure-a laser-induced breakdown(LIB)-can be used to disintegrate urinary calculi.
A high intensity Nd-YAG laser was developed at our hospital-in cooperation with the Department of Experimental Physics,Karl-Franzens Universität in Graz-for intracorporeal laser-induced shock wave lithotripsy of ureteral- and kidney calculi.

Physical and technical aspects of laser-induced shock wave lithotripsy (LISL)

Laser induced shock waves with an extremely steep shock wave front and a high pressure amplitude are generated by localized optomechanic energy conversion from a Q-switched,nanosecond pulsed Nd-YAG laser (2,5,6)

At the interface between the surface of the calculus and the surrounding liquid an electrical breakdown-laser induced breakdown(LIB)-is created by increasing the power density of the laser beam.In the focus part of the fluid vaporizes and a tiny plasma filled bubble is generated.By expansion and attentuation of this localized plasma,a shock wave front is emitted and propagated in the medium.Expansion and cooling of the plasma results in an oscillating plasma bubble, which causes cavitation in a liquid.With a Nd-YAG laser of 1064nm wave length,8nsec pulse duration and a single pulse energy of 20-80mJ,shock waves can be generated with peak pressures of 1000 bar in less than 4nsec.The medium applicated energy is at maximum some Watts,so that in practice no thermic effects at all are involved.Moreover the interval between two short term pulses is so high(50 Hz pulse repetition rate-8nsec pulse duration), that thermic effects due to a series of pulses can be ruled out.The ratio of a nanosecond pulse ($8x10^{-9}$sec) to 50Hz repetition rate(50pulses sec)-as applied in our laser system-would equal in a more comprehensive enlarged scale a pulse of 1 second duration followed by a pause of 28.9 days.This evident comparison enligthens the laser principle using an extemely short pulse with a steep,high intensity shock wave followed by a comparably long pause.

Coupling of the Q-switched laser energy into a 600-,400-or 200μm quartz fibre was performed by a specially designed tube into the plane end of the fibre.Focussing of the laser pulses at the fibre tip in the form of a laser cone was achieved by a specially formed fibre end.

Optimal stone disintegration is performed within the laser induced break down seen at the fibre tip as a light cone.Fragmentation also can be heard as a series of slight clicking sounds.Short contact of the fibre with the stone or the spray of fragments does not damage the fibre,while constant laser irradiation in stone contact results in cracks of the fibre surface.The highly flexible quartz fibre can be easily changed during the operation by removing the plug-formed end from the coupling device in the laser.

Biologic effects of laser irradiation

Biologic effects of laser irradiation were evaluated by irradiation of cell cultures,whole blood and urothelium of bladder,ureter and kidney parenchyma in pigs.Laser light from a Nd-YAG laser(λ=1064nm,8nsec pulse duration)was irradiated either as a single pulse or a pulse series of 20Hz.Single pulse energy ranged between 50-120mJ and was up to four times higher than it is now used in patient treatment(35mJ).

No macroscopic alteration could be seen on the pig urothelium 2,4,8

and 12 days after laser irradiation.5µmserial cuts showed no histologic change,especially no thermic damage,necrosis or hemorrhage.Electronmicroscopy also did not reveal any tissue alteration.
Biologic effects on the urothelium are limited to a tiny mechanic rupture(100µm width,40µm depth)of the tissue without creating thermic effects or a hole through the ureteral wall.This mechanic effect of the shock wave is still confined to the urothelial mucosa.
These results are a prerequisite for harmless application of pulsed nanosecond laser energy in the urogenital tract of man.Inadvertent irradiation of the urothelium during laser stone disintegration therefore does not cause any side effects(3,4)

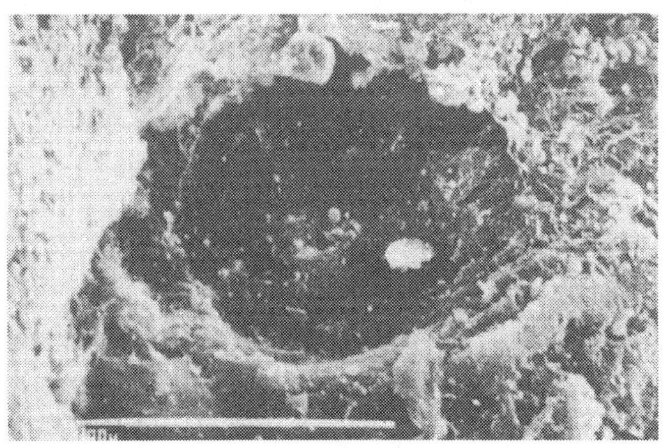

Electronmicroscopy of the ureter immediately after laser application (60mJ,20 pulses).600fold magnification.Note scale:the 'cone' is about 100µm wide and 40µm deep.

Clinical experience with laser-induced shock wave lithotripsy

From June to December 1987 41 patients with a total of 43 urinary calculi have been treated.Only stones,which did not pass spontaneously out of the ureter after at least four weeks conservative treatment,obstructive ureteric stones and ureteral-and kidney stones not suitable for ESWL treatment were selected.27 of 39 ureteral stones were obstructing the ureter with consecutive upper urinary tract dilatation.28 patients had general anaesthesia,9 ureteric calculi and all four kidney stones were fragmented in local anaesthesia.
A 11.6F rigid ureteroscope was inserted into the bladder and under maximum inflow of the irrigation solution gently advanced directly into the ureter in 10 patients.In another 29 patients,a flexible 4F ureteral catheter was passed into the ureteral orifice and to the calculus.With a slight torsion of the whole instrument using the ureteral stent as a protection of the roof of the orifice and as a guide to the

stone,the ureteroscope was gently advanced to the calculus.No dilation of the orifice was necessary.The quartz fibre was directly inserted into the ureteroscope (35 patients).Laser stone disintegration is done under constant vision using an irrigant solution.The calculus is brought into the focus of the laser by the help of a red Helium-Neon pilot laser.

For kidney stone fragmentation,the kidney pelvis is punctured under fluoroscopic control,the channel dilated to 26F,the nephroscope advanced and the laser fibre directly pushed to the calculus.

In 6 cases the fibre was passed through a 700μm rigid metal channel fixed to another 500μm channel used as a suction tube.This irrigation circuit with drainage of the tiny stone particles(size less than 0.5mm) was considered unnecessary.Stone disintegration under constant vision using only inflow of an irrigation solution was sufficient.Tiny stone particles or blurring of the fluid could be flushed away by constant inflow or drained out through the outlet of the ureteroscope.

In 31 patients(33 calculi),the stone was fragmented into tiny particles or stone powder and flushed out through the ureteroscope.In 7 patients the stone was reduced to a size small enough to be extracted by forceps together with the ureteroscope.Three stones of hard amorphous calcium-oxalate-monohydrate were too hard to be completely disintegrated by the laser in reasonable time.Those calculi were further fragmented by ultrasound after a considerably long time (30min)and stone parts removed by basket or forceps(1 case) or flushed up to the kidney and taken out percutaneously(2 cases).

Laser stone disintegration was performed within 20seconds to 5 minutes irradiation time(1000-15000 pulses) with an average time of 22.8 minutes for the whole operation.No harm was done to the ureteral wall by inadvertent laser irradiation to the urothelium(4 cases) or the manipulation with the ureteroscope or the laser fibre itself.

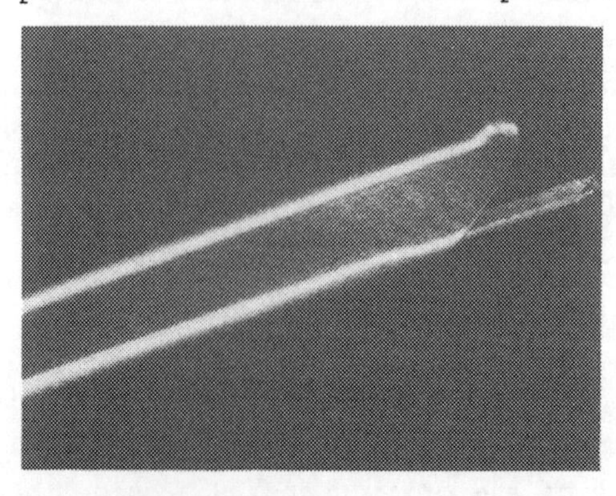

Tip of the rigid ureteroscope with 600μm quartz fibre(core diameter).

All patients were free of stones immediately following laser stone disintegration, proven by plain X-ray film postoperatively and an IVP some weeks later.No additional methods as removal of stone fragments by basket or loop were required.No residual stones were left in the ureter or kidney in order to pass spontaneously.Laser stone fragmentation proved to be effective in all but three hard calcium-oxalate-monohydrate stones.Optimal conditions for stone fragmentation with the Nd-YAG laser of 1064nm wavelength were recorded with 8nsec pulse duration,35-50mJ pulse energy at the fibre tip and 40-50 Hz repetition rate(table 1).

Table 1

31 patients(33stones)	complete stone disintegration
7 patients	reduction of stone size,then removal by forceps together with the endoscope
3 patients	fragmentation impossible or too slow(ca-ox-monohydrate)

stone analysis: calcium-oxalate-dihydrate(n=17)
calcium-oxalate-monohydrate(n=16)
struvite (n=4)
uric acid (n=7)

time of stone fragmentation:20 sec to 5 min(1000-15000 pulses)
total average operation time:22.8 min

Irrigation during stone disintegration is essential,as the spray of fragments impairs laser fragmentation by energy absorption.For completely obstructing ureteric stones a good drainage of the irrigation fluid is mandatory thus providing a clear liquid between the stone and the laser fibre.The focussed shock waves of the laser do not propulse the stone in the ureter,however vigorous inflow of the irrgiation saline can flush the already diminished calculus up in the ureter. Thereforeirrigation has to be reduced at the end of the operation or laser application only be performed with outflow of the fluid.In 7 patients, who had bigger obstructing stones with grossly enlarged ureter and kidney pelvis above the stone,the calculus was gently advanced by the instrument,the stone passed by a stent with a Dormia basket and fixed within the basket.Thereafter laser stone fragmentation was performed until the stone was completely disintegrated (4 patients) or the stone was fragmented until it was small enough to be easily removed within the basket. Three obstructing stones even had to be partly disintegrated by the laser for passage of a Dormia basket afterwards.

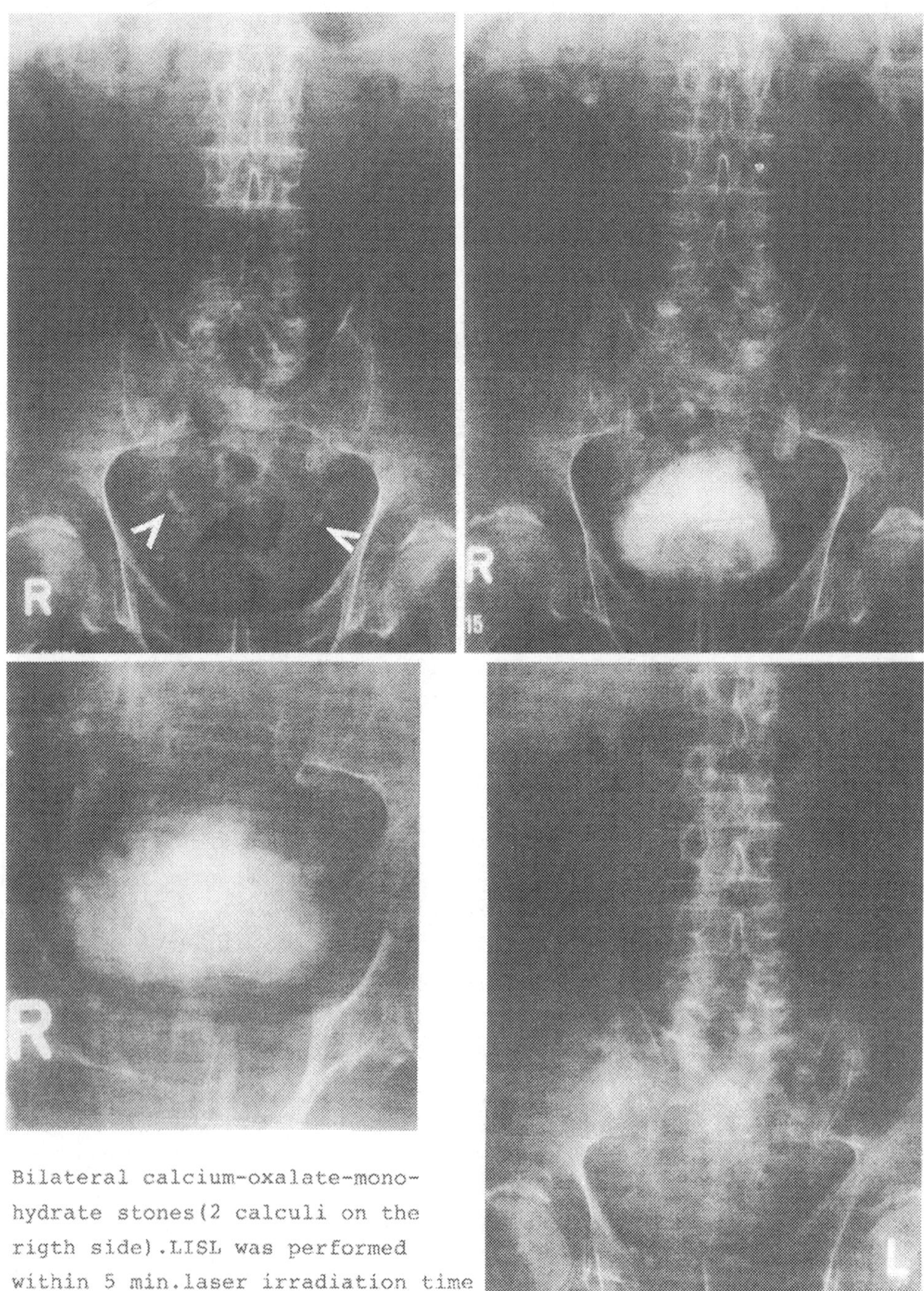

Bilateral calcium-oxalate-monohydrate stones(2 calculi on the rigth side).LISL was performed within 5 min.laser irradiation time Patient was free of stones after laser stone disintegration and drainage of stone parts out of the ureter

So far stone disintegration was only performed with rigid ureteroscopes. The flexible quartz fibre(0.2,0.4 or 0.6mm) however allows miniaturisation of these instruments and application with small flexible probes. The laser action allows controlled fragmentation of the calculus, so that no additional measures as removal of stone parts with forceps or basket are necessary.Bigger stone particles created at the disintegration site with electrohydraulic or ultrasonic probes have to be collected from the ureter resulting in prolonged manipulation and stress to the ureteral orifice.Laser stone disintegration proved to be a 'one-step' procedure in the ureter, as no removal of the instrument out of the orifice is necessary.

Effective and secure stone disintegration with small flexible fibres is a prerequisite for application with flexible endoscopes,so far only used as diagnostic instruments.Using steerable, flexible probes or even laser irradiation under fluoroscopic control with the laser fibre alone or with the help of our newly designed laser basket for application with the YAG laser(0.2mm fibre), stone disintegration is even less invasive and can be easily done in local anaesthesia.

The spectrum of indications for laser stone disintegration using active flexible endoscopes consists of caliceal stones not accessible by rigid percutaneous renoscopy,all ureteral calculi,especially those not suitable for ESWL therapy or disintegration of Steinstraße following ESWL.

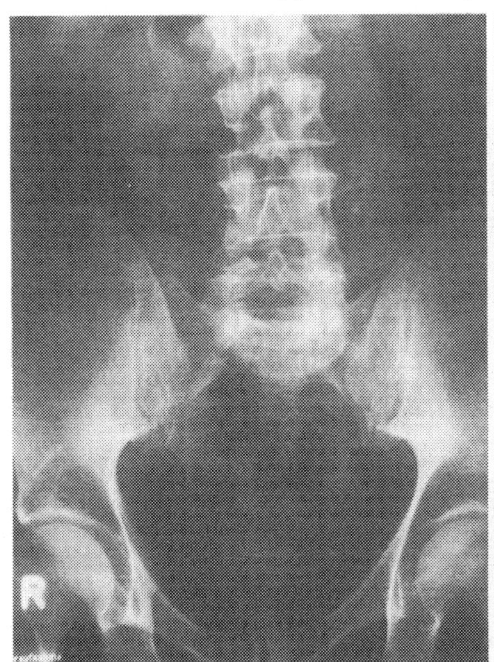

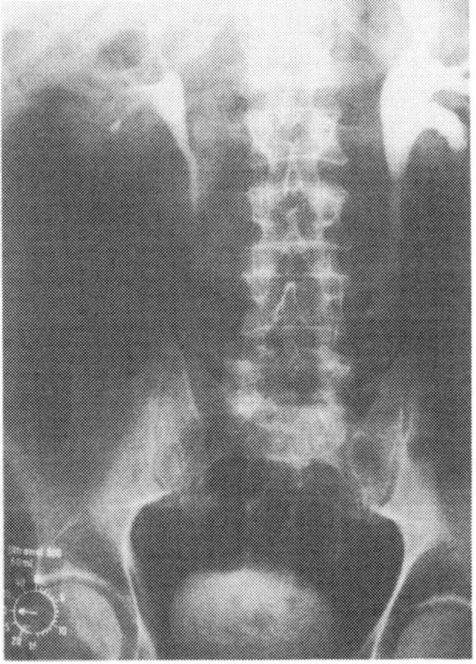

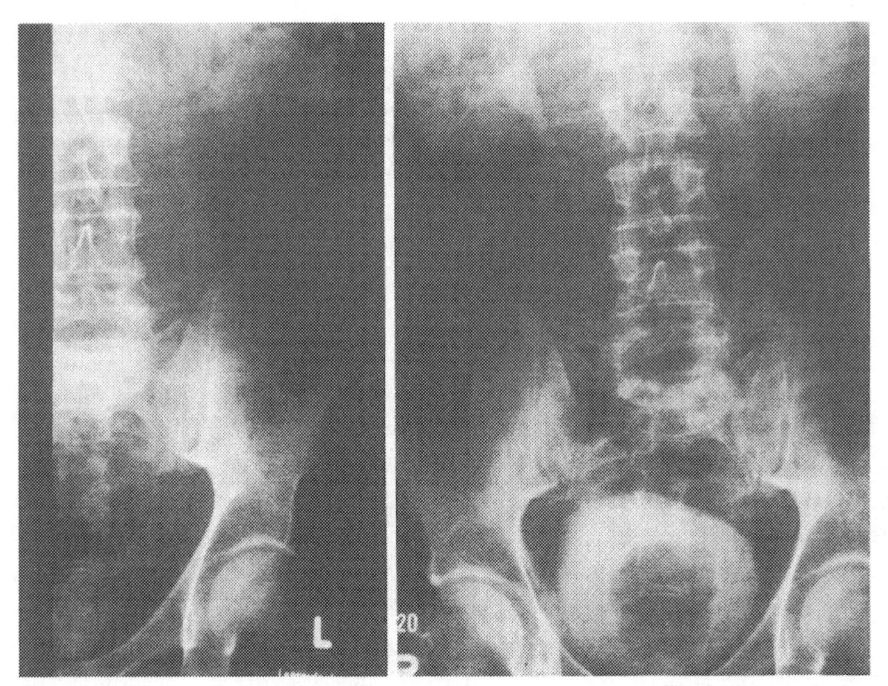

Obstructing calcium-oxalate-monohydrate stone .Complete disintegration with LISL (3 min irradiation time, 50 Hz rep.rate and 50 mJ pulse energy IVP 3 weeks later showing patent left ureter without hydronephrosis.

1.Hofmann R.,Schütz W.:Zerstörung von Harnsteinen durch Laserstrahlung Urologe A 23,181-184 (1984)

2.Hofmann R.,Hartung R.,Schmidt-Kloiber H.,Reichel E.:Laser-induced shock wave lithotripsy(LISL)-biologic effects and first clinical application, Proc 7th Congr.Int.Soc.Laser Surg.and Med.,Munich 6/87

3.Hofmann R.,Hartung R.,Schmidt-Kloiber H.,Reichel E.,Schöffmann H.: Morphologische Untersuchungen des Urothels nach Einwirkung intensiver nanosekunden Laserpulse,Urol Int.(1988) in press

4.Hofmann R.,Hartung R.,Schmidt-Kloiber H.,Reichel E.,Schöffmann H.: laser-induced shock wave lithotripsy- biologic effects of nanosecond pulses, J.Urol in press

5.Schmidt-Kloiber,Reichel E.,Schöffmann H.:Laser-induced shock wave lithotripsy(LISL), Biomed Technik 30,173 (1985)

6.Schmidt-Kloiber H.,Reichel E.:Die Abhängigkeit der Druckamplitude einer Stoßwelle von der Feldstärke beim laserinduzierten Durchbruch in Flüssigkeiten, Acustica 54,284-288 (1984)

Dr.Rainer Hofmann
Dep.of Urology
Technische Universität München,Klinikum re.d.Isar
8000 Munich 80
FRG

The Pulsed Dye Laser in Clinical Stone Problems

G.M. Watson

The Middlesex Hospital, Department of Urology,
Mortimer Street, London WIN 8AA, United Kingdom

Summary

The pulsed dye laser has been used to fragment ureteric calculi in 250 patients. Its contribution to the Urologist's armamentarium lies partly in that it is safer to use as a fragmentation modality than the electrohydraulic probe and partly in the advantages it offers in ureteroscopy. As a result of the ability to use smaller calibre ureteroscopes with the laser ureteroscopy has become easier and the complications of stricture, failure to reach the stone and ureteric perforation have diminished. The future development of this technique may be by fragmenting the calculus without endoscopic control.

Introduction

A pulsed dye laser when optimised for stone fragmentation has a pulse duration of 1 microsecond, a wavelength of 445 nm and is delivered through a 200 micron fibre WATSON et al. [1]. These parameters were modified for clinical application by selecting a wavelength of 504 nm where the differential of absorption by the stone over the ureter is maximal. This system was assessed in the pig ureter before being applied clinically WATSON et al. [2]. In this animal study human urinary calculi were impacted into the proximal ureters of pigs and fragmented using the laser in one ureter and the electrohydraulic probe in the other. The fragmentation was performed under vision using ureteroscopes of 10 and 12 F calibre. The conclusions of this study were that the laser caused negligible or no damage at the site of fragmentation in stark contrast to the electrohydraulic probe. The distal ureter was more prone to injury than was the proximal ureter and it was possible to distinguish the injury resulting from ureteroscopy from the local damage from stone fragmentation. There was significantly more injury resulting from insertion of a 12 F ureteroscope than from insertion of a 10 F ureteroscope. Thus the main contribution of using a laser system for ureteric stone management lies as much or more in its implications for endoscopy as in its action as a lithotripter.

This paper describes the clinical results using the Candela pulsed dye laser from October 1985 until recently. The development of the technique has been principally in the development of the endoscopy.

Patients and Methods

This series describes a consecutive series of patients with ureteric stones referred to one consultant at The Institute of Urology. These patients were mainly referred from other urologists and 15% had already failed some previous instrumentation attempt. There were also 18 patients with obstruction following Extracorporeal Shock Wave Lithotripsy (ESWL). The age range of the patients was 12 to 87 (mean 52). The stones ranged in width from 5 to 18 mm (mean 7.9 mm) and from 2 to 27 mm in length (mean 9.3 mm). The stones had been present in the same position in the ureter as proven on Xray for between 3 days and 32 months (mean duration 6 months).

In addition to the ureteric calculi 5 children have been treated with vesical calculi. These children all had reconstructed urethras and bladder necks which limited the size of cystoscope which could be inserted. The laser fibre was passed through a 10 F cystoscope for these treatments. The majority of procedures have been conducted under endoscopic control. The details of the endoscopes has evolved and will be discussed in each subseries. When performing a procedure under vision the fibre was advanced until seen to be in contact with the stone. A series of bursts of pulses at 10 Hz were delivered to the stone surface. The pulse energy was adjusted according to the effect. The end point was when all fragments were less than 2 mm in any diameter.

Results

1) The first 100 patients

The ureteroscope of first choice for this series was the 11.5 F short Wolf with a conventional full length ureteroscope inserted for stones out of range with the shorter version. The instrument channel was 6F which was large in comparison to the laser fibre. Therefore the laser fibre was introduced through a catheter so that irrigant could flow back through the catheter around the laser fibre. In this way flow rates of up to 120 ml/minute were achieved with improvement in visualisation. The ureteroscope was introduced in the conventional retrograde manner in 95 of the patients and antegradely (for upper third of the ureter calculi) in the remaining 5. The ureteric orifice required dilatation to 12 F in 65 of the 95 retrograde ureteroscopies. The 200 micron fibre was used with pulse energies of 25 mJ increased as necessary to 50 mJ at a pulse repetition rate of 10 Hz.

There were 21 patients in whom the ureteroscope could not be passed to the level of the stone. Where possible a double J stent was passed alongside the stone and a second attempt was made at ureteroscopy after an interval of 1 to 4 weeks. 20 of these patients elected to have a second attempt at ureteroscopy and one elected for an open operation. The stone was successfully fragmented in all of these 20 at a second attempt. In 18 patients either the whole stone or a significant fragment from it was flushed with the irrigant back into the renal pelvis. In these cases the ureter was stented and a percutaneous treatment or ESWL performed. There were 6 patients in whom the ureter was perforated during the attempt at ureteroscopy before reaching the stone. Either a ureteric stent or a percutaneous nephrostomy drain was required.

The stone was reached eventually in 99 of these patients. There was one patient in whom the stone could not be fragmented (principally a calcium oxalate monohydrate stone). Electrohydraulic lithotripsy also failed and the stone was removed by open operation. In each of the remaining 98 patients the stone or stones were fragmented effectively without any sign of injury to the surrounding ureter. There was a clear acoustic feedback to the operator when the pulses were absorbed by the stone. Once this feedback was lost the operator discontinued the train of laser pulses and repositioned the fibre tip on the stone. The time taken to fragment the calculi was dependent on the energy used and the chemical composition. The majority of calculi were fragmented in a few minutes. Oxalate stones with a high proportion of monohydrate tended to split into a few relatively large fragments which had to be basketed out.

All patients subsequently had follow up by intravenous urography 6 months following their procedure. 2 developed significant strictures in the distal ureter, one of whom required reimplantation of the ureter. These strictures of the lower third of the ureter were both in cases of upper third stone fragmentation. Even though the injury to the ureter was clearly a result of ureteroscopy there was no injury to the ureter noted at the time of the endoscopy. There were no strictures at the site of stone fragmentation.

The complications varied according to the level of the stone in the ureter. There were 40 patients in whom only the lower third of the ureter was instrumented. Of these 40 there were two failures to reach the stone at the first attempt and no cases of perforation or stricture. There were 28 patients requiring ureteroscopy to the middle third. There were 8 failures to reach the stone on the first attempt and there were 2 perforations of the ureter. There were 32 patients with stones in the upper third of the ureter. There were 11 failures to reach the stone on the first attempt, 4 perforations of the ureter and 2 subsequent strictures of the lower ureter. There were no complications and no failures to reach the stone in the 5 patients in this group who had an antegrade ureteroscopic approach.

2) The second 100 patients

The following modifications were made early in this second series. The fibre was altered to a 320 micron diameter so that although there was a slight rise in the threshold for the majority of calculi the calcium oxalate monohydrate calculi could be fragmented efficiently. The second modification was that the 8.5 F Wolf and the 7.5 F Reichert ureteroscopes became available. The 8.5 F Wolf is now the instrument of first choice. It can be introduced into the distal ureter without dilatation of the ureteric orifice. The instrument is less rigid than the larger calibre ureteroscopes and occasionally a larger calibre endoscope has to be passed in order to negotiate the iliac vessels and pelvic brim.

Using the slightly larger diameter 320 micron fibre there were no further cases of calculi which could not be fragmented. This fibre is still narrow enough to make no significant change to the flow of irrigant through the common instrument and irrigation channel. It has the advantage of being slightly more rigid than the 200 micron fibre and is easier to manoeuvre onto the stone surface. There have been no cases of failure to reach the stone on the first attempt with this increased range of ureteroscopes. There has only been one ureteric perforation (for an upper ureteric stone) which responded to stenting. There has been no case of ureteric stricture.

Thus the smaller instrumentation made possible by the small size of the laser fibre has had a significant effect on the success rate and the safety of ureteroscopy as seen in the contrast between the first and second series of 100 patients.

3) Using the laser without endoscopy

The laser was inserted through a simple 6 F catheter which had been passed under Xray control to a point just beneath the stone. Bursts of pulses of 30 mJ (via a 200 micron fibre) were delivered to the region of the stone while listening for the acoustic feedback of the plasma using a stethoscope on the patient's abdominal wall. 11 patients have been treated in this way. In only 5 has there been any acoustic signal suggesting plasma formation and in one of these the signal was only present when there was significant haematuria. There was a degree of fragmentation in 4. 2 patients passed their stone fragments shortly thereafter. In one patient the stone fragments were flushed back into the kidney and required ESWL. The fourth patient had a conventional ureteroscopy and further laser fragmentation. In the remaining 6 patients in whom there was no acoustic signal there was no evidence of having made contact with the stone and they were treated by laser fragmentation under endoscopic control.

Discussion

Just as the pig ureter model had shown that ureteroscopy is a more significant cause of trauma than the method of fragmentation so this paper has demonstrated a striking reduction in morbidity when using smaller calibre ureteroscopes. The laser fibre was significantly less damaging than the electrohydraulic probe in animal studies and could even be used directly on the ureter wall without causing significant damage. However it is the ability to use the laser via smaller calibre endoscopes which has been the most significant advance. Ureteroscopy becomes easier to perform because there is no need to dilate the ureteric orifice and there is no resistance to the advance of the endoscope as it ascends. The problem area remaining is the negotiation of the iliac vessels and pelvic brim. Small calibre ureteroscopes are prone to deform and the optics become distorted with "half moon" views. The use of a laser via a catheter under Xray control would be an important advance. The further development of this mode of treatment now requires purpose-built catheters to direct the fibre away from the ureteric wall. The use of plasma emission spectroscopy for feedback of contact with the calculus will also be valuable.

References

[1] The pulsed dye laser for fragmenting urinary calculi. G.M.Watson, S.Murray, S.P.Dretler, J.A.Parrish. Journal of Urology (1987) 138: 195-198.

[2] An assessment of the pulsed dye laser for fragmenting calculi in the pig ureter. G.M.Watson, S.Murray, S.P.Dretler, J.A.Parrish. Journal of Urology (1987) 138: 199-202.

Gallstone Lithotripsy by Pulsed Nd:YAG Laser

H. Wenk[1], *V. Lange*[1], *K.O. Möller*[1], *F.W. Schildberg*[1], and *A. Hofstetter*[2]

[1]Department of Surgery, University of Lübeck,
Ratzeburger Allee 160, D-2400 Lübeck, Fed. Rep. of Germany
[2]Medical Laser Center, University of Lübeck,
Ratzeburger Allee 160, D-2400 Lübeck, Fed. Rep. of Germany

Introduction

Continuous wave Nd:YAG lasers are used in curative and palliative surgery. The development of a pulsed Nd:YAG laser now also allows the desintegration of concrements (1,4). The presupposition for clinical use of such technique is a system that produces the necessary energy in very short pulses and high repetition rate to avoid thermal lesions and flexible optical transmission systems that can be guided through the working chanel of flexible endoscopes.

Experiments

1. in vitro experiments

For our in vitro experiments a pulsed Nd:YAG laser with a wave length of 1.064 nm and a pulse duration of 12 nsec was used. The frequency was variable: the examinations were done with pulse energies of 45 mJ and frequencies of 20 cycles/sec. Two different opto-mechanical couplers were used.

The first transmission system was a flexible quartz fiber with a diameter of 1.000 micrometer. At the distal end a lense system was attached to focus the emitted laser light in the frontal plane. The lense system used had a diameter of 4,5 mm and was therefore not suitable for endoscopical use.

This was the reason for the development of substantially smaller optomechnical coupler. It's quartz light-guide has a diameter of 600 micrometer and is implanted concentrically into a rinsing tube. The optomechanical coupler at the distal end has a diameter of 2,2 mm and can be used in modern flexible endoscopes.

All gallstones were measured in 3 planes and also weighted. The stones could be destroyed in short time. Stones with a weight of up to 1/2 gramme were destroyed within 5 minutes. Only one stone required more than 15 minutes for total desintegration.

Gallstone-Lithotripsy by pulsed Nd:YAG-Laser

	I	II	III	IV	V	VI	VII	VIII	IX	X	XI
vol. (ml)	468	1300	336	510	566	353	242	303	392	497	
weight (mg)	268	880	210	270	355	218	170	179	190	458	5300
main const.	Chol	Chol	Chol	Chol	Chol	Chol	Chol	Chol	Chol	Chol	Bili
duration	2'15"	15'33"	4'50"	3'	4'50"	3'20"	2'5"	4'	3'35"	5'30"	3'50"

The duration of lithotripsy depended on size and weight of the gallstones and on the other hand on their hardness. The systems, consisting of lense systems and optomechanical coupler, showed no differences in quality of lithotripsy.

2. Animal experiments

To examine the soft tissue reactions on the bile duct and the gall bladder we carried out experiments in 8 pigs with a weight of 30-40 kg. After laparotomy of the upper abdomen the main bile duct and the gall bladder were exposed and the bile duct and the gall bladder were opened between sutures over 2-3 cm.
Different tissue areas were exposed to the laser pulses over 5 and 15 minutes.
The withdrawl of the areas was done immediately, in other experiments after one hour, 3 days, 2 weeks, 4 weeks and 6 weeks. Photos were taken from the preparations, subsequently they were fixed in formaline and given to the histological processing.

The results: severe defects could not be observed. There were no perforations by the treatment, neither coagulation of the tissue nor stenosis. The only tissue reaction were foreign body cells in the regions of the suture markings.

LASER INDUCED SHOCKWAVE LITHOTRIPSY
TISSUE REACTION

		immediately	1 hour	2 weeks	6 weeks
5 min	macro	edema	edema	NAD	NAD
	micro	edema	edema	NAD	NAD
15 min	macro	haematoma	haematoma	NAD	NAD
	micro	bleeding into submucosa	bleeding into submucosa	NAD	NAD

Evident was an edema of the bile duct's wall and the gall bladder immediately after shock wave lithotripsy. After a treatment ofer 15 minutes petechial bleedings into the bile duct's wall were recognized. They corresponded to haematomas in the submucosa in the microscopical picture.

The tissue which was explanted one hour after laser exposition showed supplementary a leucocyt diapedisis.

The reactions of the bile duct's wall were totally reversible and couldn't be recognized in the late explantated bile ducts.

Clinical application

Beneath the endoscopic application there is an indication for shock-wave-lithotripsy also in intraoperative situations, if it is possible to reduce the dimension of the operation (3).

An example: a 40 years old female patient had pain in the stomach since her 17th birthday. Diagnostic procedere in 1985 showed a chronic calcifying pancreatitis as the cause of pain. Endoscopic retrograde cholangiopancreaticography showed an incarcerated concrement in the head of the pancreas.

By acceleration of pain the operation was indicated and a pancreaticojejunostomy was planned. After incision of the pancreatic duct which was dilated, multiple concrements could be extracted.The big stone in the head of the pancreas could neither be exposed nor be removed.

Intraoperative endoscopy showed the concrement fixed in the small ducts. By endoscopically applied laser induced shock-wave-lithotripsy it was possible to destroy the concrement completely and to remove the fragments.

Afterwards the papilla of vater and the terminal bile duct could be inspected without difficulties.

The intervention could be finished as a pancreaticojejunostomy in favour of a duodenopancreatectomy.

The 40 years old patient was better attended by a draining operation, for resection of the pancreas is incriminated by endocrine insufficiency, diabetes and higher letality in comparison to spontanous progress (2).
The draining operation was made possible by laser lithotripsy of the impacted stone.

Summarizing, we see indications for laser induced shock-wave-lithotripsy in impacted stones of the common bile and pancreatic duct that cannot be removed by any other methods. By means of the endoscopic intraoperative application the dimension of surgical intervention can be significantly reduced.

Literature

1. **Hofstetter,A., Schmeller,N., Pensel,J., Arnholdt,H., Frank,F., Wondrazek, F.:**
   Harnstein-Lithotripsie mit laserinduzierten Stoßwellen.
   Fortschr.Med., 104, 654, 1986

2. **Horn,J.:**
   Therapie der chronischen Pancreatitis.
   Springer Berlin,Heidelberg, New York, (1985)

3. **Schildberg,F.W., Lange,V., Wenk,H., Schüller,J.:**
   Die intraoperative, endoskopische Lithotripsie von Pankreaskonkrementen.
   Chirurg 58, 239-242 (1987)

4. **Schmeller,N., Baumüller,A., Hofstetter,A.,:**
   Nichtoperative Behandlung von Harnleitersteinen mit Hilfe der Ureterorenoskopie.
   Fortschr.Med., 102, 895, (1984)

Electronmicroscopic Evaluation of Urinary Stones Following Laser Stone Disintegration in Patients

R. Hofmann[1], *R. Hartung*[1], *H. Schmidt-Kloiber*[2], *E. Reichel*[2], *and H. Schöffmann*[2]

[1]Department of Urology, Technische Universität München, Klinikum rechts der Isar, D-8000 München 80, Fed. Rep. of Germany
[2]Department of Experimental Physics, Karl-Franzens-University, A-8010 Graz, Austria

Intracorporeal laser-induced shockwave lithotripsy(LISL)with a high-intensity Nd-YAG laser has proved to be an effective and secure procedure for the treatment of ureteral calculi.Especially bigger,obstructing stones can be completely fragmented into very tiny particles or stone powder.The fragments are flushed out through the endoscope during laser treatment, so that the patient is free of stones immediately following the operation.A high intensity Nd-YAG laser with nanosecond pulses can generate a shockwave in a liquid suurounding a stone. Shock waves reaching an area withdifferent sound wave impedance-e.g. urinary calculi-are partly reflected.At the front of the calculus compressive and at the rear tensile pressure is generated with consecutive disintegration of the stone.The laser beam is coupled into small,highly flexible quartz fibres(0.2,0.4 and 0.6mm core diameter). The laser energy is focussed at the fibre tip by a specially shaped end. During laser application,the laser-induced breakdown(LIB) can be seen as a light cone beginning at the fibre tip.Stone disintegration can be done within the LIB.Optimal fragmentation can be heard as a series of slight clicking sounds.(1,2).

In an experimental setting, the stone and the laser fibre is brought under water and the stone disintegrated under visual and acustic control.After a few seconds the liquid is blurred by stone powder thus absorbing the laser energy.Effectivity of stone disintegration and sight to the stone are impaired.With an irrigation of fresh saline to the stone,the liquid around the calculus is clear and LISL can be optimized.Different stone types show different hardness and resistance to laser stone disintegration.Easily fragmentable stones as apatite or struvite/apatite stones are disintegrated into stone powder, thus opacifying the irrigation fluid in greater extend than harder stones as amorphous calcium-oxalate-monohydrate or uric acid stones.

Variation of the single pulse energy and the pulse repetition rate results in a different size of stone particles.In vitro stone size and-in relation to fragmentation time -the efficacy can be easily determined by two methods:

1. the stone is weighed and then packed into a net with defined mesh width. Laser irradiation creates tiny stone particles, which fall out. The residual stone is weighed afterwards.
2. A stone is disintegrated in a certain time. Stone particles are either laid on a net or raster and stone size is measured or stone particles are filtered through nets with different mesh width and so different particle size can be evaluated.

These procedures allow correlation of the efficacy of stone disintegration (time/stone size) in relation to the composition of the stone.
In an experimental setting, all stones regardless of their composition could be easily fragmented in short time and laser parameters were optimized in this system.
In patient treatment however we realized that the experimental data for laser stone disintegration were not optimal for LISL in the ureter.
In order to improve parameters for LISL in vivo, a new method to verify the efficacy of different treatment options at the same stone was evaluated.

Material and methods: During laser stone disintegration using a rigid ureteroscope and a 600µm quartz fibre, single pulse energy and repetition rate of the laser was changed as well as the irrigation fluid. The stone particles created during the treatment were flushed out through the endoscope and collected separately. Stone analysis was performed with X-ray diffractometry. Laser stone fragmentation was thus performed always on the same stone, allowing a direct correlation between stone composition and laser parameters. Stone particles were weighed and classified according to their size. Raster electronmicroscopy was performed to evaluate particle size and form.
Three different calcium-oxalate-monohydrate and three calcium-oxalate-dihydrate stones were irradiated with 35-, 45-and 50mJ single pulse energy in the ureter. Also pulse repetition rate was changed from 30 to 50Hz and two different saline concentrations for irrigation were used (o.9% and 1.4%NaCl).

Results

Irradiation of the same stone showed, that efficacy of laser stone disintegration was increased with higher single pulse energy and higher repetition rate. Moreover stone particles created with higher energy were smaller. Higher concentrated saline also improved LISL by generating a more powerful laser induced breakdown in the liquid, thus increasing efficacy and reducing particle size.

Efficacy of LISL in patient treatment

Nd-YAG laser λ=1064nm   600µm quartz fibre
pulse duration 8nsec
calcium-oxalate-monohydrate and -dihydrate stone

	35mJ		45mJ		50mJ	
	30Hz	50Hz	30Hz	50Hz	30Hz	50Hz
0.9% NaCl	+	+	++	++	+++	+++
1.4% NaCl	+	++	+++	+++	+++	+++

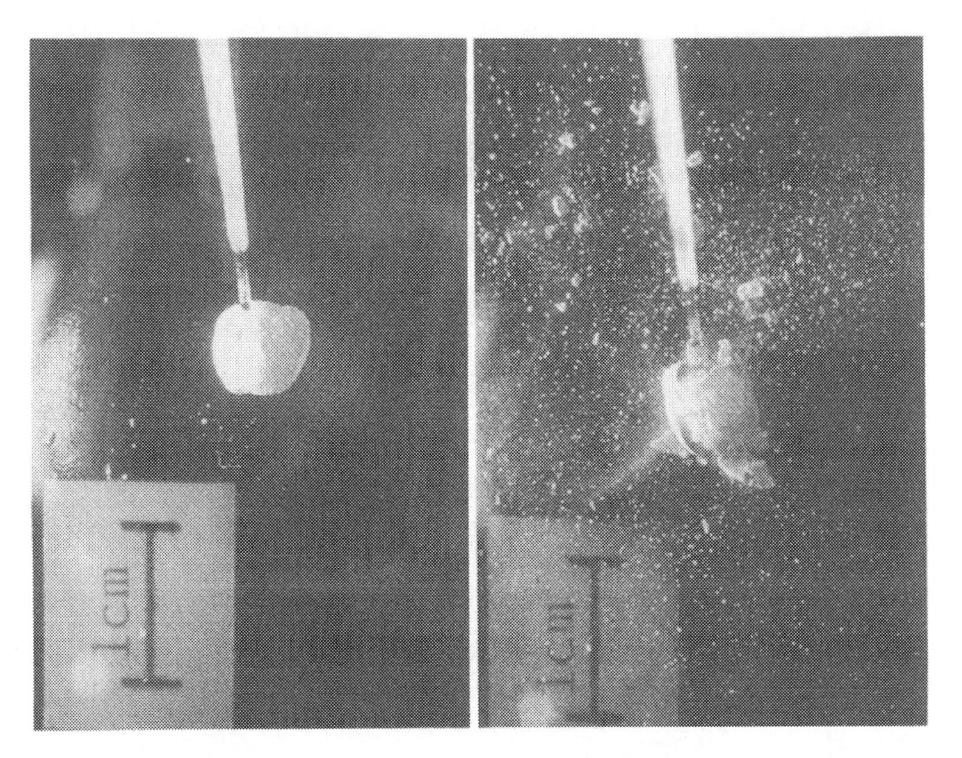

Calcium-oxalate-mono-and dihydrate stone (in vitro)

1 sec(50 pulses)with 50mJ energy at the fibre tip following LISL

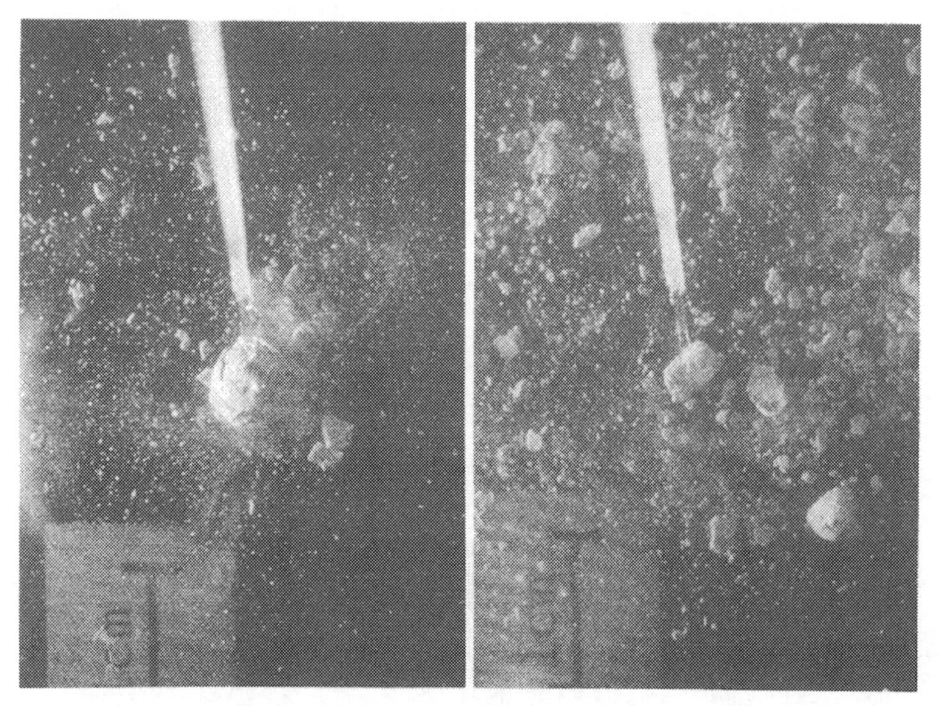

2 sec (100 pulses)following LISL 15 sec(750 pulses)following LISL

Optimal parameters for intracorporeal laser-induced shock wave lithotripsy have been found to be 8nsec pulse duration,50Hz repetition rate and 50mJ single pulse energy(600µm fibre).Constant irrigation and drainage of the stone powder is mandatory to keep vision and efficacy of the laser treatment optimal.

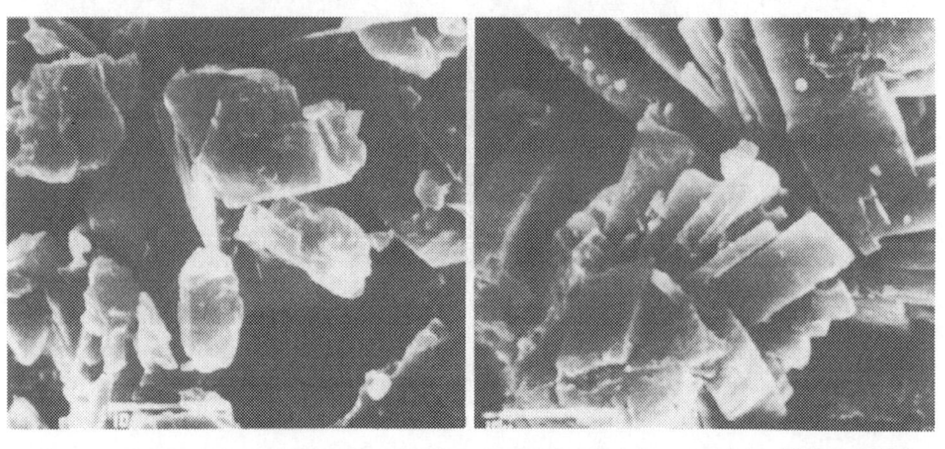

Electronmicroscopy following LISL in the ureter(Ca-ox-monohydrate, 50mJ,50Hz,1.4%NaCl)2400 fold magn.

REM of the same stone(35mJ,30Hz 0.9% NaCl).2400fold magnification Note bigger particle size.

1. Hofmann R., Hartung R., Schmidt-Kloiber H., Reichel E., SchöffmannH.: Clinical Experience with laser-induced shock wave lithotripsy Proc. 1st International Symposium on Laser Lithotripsy 10/87
2. Schmidt-Kloiber H., Reichel E., Schöffmann H.: Laser induced shock wave lithotripsy (LISL), Biomed Technik 30, 173 (1985)

Dr. Rainer Hofmann
Dep. of Urology, Technische Universität München
Klinikum re.d.Isar
8000 Munich 80
FRG

Index of Contributors